MECHANICAL BASIS
OF
BIOMECHANICS

Mechanical Basis of BIOMECHANICS

Dr. Dhananjoy Shaw

M.Sc. (Gold Medallist), M.Phil. (Specialization in Kinesiology & Biomechanics), Ph.D. (LNIPE)

Senior Lecturer,

Indira Gandhi Institute of Physical Education and Sports Sciences (University of Delhi), Delhi

SPORTS PUBLICATION

G-6, 23/23-B, EMCA House, Ansari Road,
Darya Ganj, New Delhi-110002
Ph.: (O) 65749511 (M) 9868028838
(R) 27562163. Fax: 011-23240261
E-mail: lakshaythani@hotmail.com

Published by:

SPORTS PUBLICATION
G-6, 23/23-B, EMCA House, Ansari Road, Darya Ganj, New Delhi-2
Ph. : (O) 65749511 (R) 27562163, (M) 9868028838
Fax: 011-23240261
E-mail: lakshaythani@hotmail.com

I.S.B.N. – 81-86190-64-3

PRINTED IN INDIA 2014

Laser Typeset by:
JAIN MEDIA GRAPHICS,
C-4/95-A, Keshav Puram, Delhi-110035. Ph.: 20296366, 9911151534

Printed by:
CHAWLA OFFSET PRINTERS, Delhi-110052

Price: Rs. 750/-

Preface

Biomechanics is one of the most progressing and promising subject in India. It has established in uniqueness inregard to application to human health, medical sciences, human engineering, manufacturing (from shoes to aircraft), man-machine mechanism, productivity, games and sports (from fitness to creation of new world records). The development of biomechanics in India going to put an outstanding milestone in the 21st century. India having very short history of biomechanics, some practice of biomechanics have been observed before 1980 in India with markedly deviated from the actual concept of the subject or substituted with some other title or sub-title. Later since the beginning of the 1980's the biomechanics have been placed in the right perspectives and have been incorporated as independent subject in various disciplines in the undergraduate, postgraduate levels and doctorate research programmes. In the year 1988, the Indian Society of Biomechanics (ISB) has been finally founded with the assistance of collaborative experts, researchers from medical, engineering and other subjects area including physical education with the kind motivation and interaction of the author's Ph.D. supervisor Prof. K.B. Sahay, IIT Delhi, as Founder General Secretary of ISB, so far the society has conducted six National Conferences on Biomechanics with great success.

It has been observed that the biomechanics and kinesiology related to human motion and its teaching with specific reference to physical education and sports have been greatly ignored in India. The reason may be the lack of expertise, misinterpretation of the subject, lack of practice and/or lack of literature. No book has been written on the subject by the Indian author, specially catering to the requirements of physical education and sports sciences. Most of the available published books are imported from outside of the

India and are very costly, not easily available to common readers/ students and at the same time those literatures are not suitable or not much friendly to the people of physical education profession, whether they are teacher, researcher or students in India.

Hence, an attempt has been made to write a book entitled **"Mechanical Basis of Biomechanics"** using simple language. Every attempt has been made to incorporate living reference model and examples common to Indian to enable the readers to understand the concepts.

The first book written by the author himself (myself) entitled **"Biomechanics and Kinesiology of Human Motion"** followed by **"Pedagogic Kinesiology"** and thereafter, ten more books motivated the author to enrich the Biomechanics by writing a book emphasising on Mechanical Basis of Biomechanics covering the theoretical and practical applications, which are useful for undergraduate and post-graduate learning as well as for readers to develop the computing abilities in Biomechanics. This book attempts to equip the readers for research applications. Thus, this book will definitely be considered as reader-friendly book enabling the beginners to lay a strong foundation for professional progress in the future prospects in the profession as well as refreshing the experts in the profession.

Author

This Book is

Dedicated

To

MAA TARA

लक्ष्मीबाई राष्ट्रीय शारीरिक शिक्षा संस्थान, ग्वालियर
(सम विश्वविद्यालय)
LAKSHMIBAI NATIONAL INSTITUTE OF PHYSICAL EDUCATION, GWALIOR
(DEEMED UNIVERSITY)

संदर्भ क्र. CD/222/99 दिनांक 27/12/99

FOREWORD

In the past few years fervent efforts have been made by various physical educationists to enhance the field of Bio-Mechanics. Following the league, visualizing the endeavor of **Dr. Dhananjoy Shaw**, a persistent effort in the field of Physical Education has been made. With his 13th book **"Mechanical Basis of Bio-Mechanics"** completed I congratulate him for bringing forth his scientifically based ideas into concrete shape making the book extremely useful for students, teachers, researchers and scientists of various disciplines.

The book has an excellent appeal because of its lucid and easily understood style emphasizing on mechanical aspects of Bio-mechanics considering both the theoretical and practical skills, as well as, research applicable to human motion.

The theoretical and practical skills are rightly interwoven by suitable technical illustrations and examples at various places transforming it into a much sought after, book for a knowledge seeker.

Considering the substance and style of the book, I recommend the book to all those, attached to the field of Bio-mechanics.

(Dr. Jitendra Singh Naruka)
DIRECTOR

SHAKTI NAGAR, GWALIOR-474002 M.P., Gram : PHYSICAL, Ph. : (O) 500917, 500918, Fax : 0751-340553, E-mail : director@gwrl.dot.net.in

Chapter-1

INTRODUCTION

Biomechanical Analysis

Bio-Mechanical analysis is the process of identifying the internal as well as the external forces acting on human body during the execution of a movement and the effect or effects produced by such forces.

Internal Force

Viscosity of muscles, angle of pull etc.

External Force

Movement force, friction, water resistance, air resistance, weight, gravitational pull etc.

Movement Analysis

Movement

- A change involving location or position or a change in posture.

Analysis

- It is the resulation of a compound object or a given movement into it's constituents element or component parts. e.g., chemical components of a compound.

Fundamental Movement

Basic fundamental movement	**Technique of sports and games**
The skill developed from years, skill required for daily life activity, e.g., walking, standing, sitting etc.	Required for games e.g. Hockey, Dribbling, hitting, passing etc.

Kinesiology

It is the science of human motion or locomotion.

Concerns with (a) Muscular area
(b) Skeletal or anatomical area
(c) Identification of type of lever in action.

Kinesiological Analysis

It is the process of identifying the joints. Muscles and the bone levers used in a skill or in a movement as well as sequence and degree of their involvement.

Mechanics

It is the branch of physics which deals with the science of motion and force.

Scope and Divisions of the Subject Mechanics

Mechanics is the science concerning the conditions of rest or motion of the object around us.

It is that branch of the science subject which deals with bodies at rest when acted on by forces, or more properly, with the relations between the forces which acting on a body keep it at rest, is called **Statics.**

Dynamics is that branch of the subject which deals with the motions of bodies under the action of forces.

This is again subdivided into two parts:

The first part called **Kinematics** deals with the different types of motions possible for a body, and the effects there of on its positions, in fact, the geometry of the motions, without enquiring into the causes which produce these motions.

The second part investigates the laws of motions, that is the relations existing between the forces acting on a body, and the motions produced thereby, and is known as **Kinetics.**

Motions

Motion is change of position. When a body is changing its position, we say it is *motion.* When it is not changing its position, it is said to be at rest.

Mechanical Analysis

It is the process of identifying of those principles and laws of "Mechanics", which are applicable in execution of a given movement e.g. Mechanics of walking.

Need of Mechanical Analysis or Biomechanical Analysis

1. Mechanical Analysis / bio-mechanics forms the basis of effecient structure of the competitive performance. e.g. performance of Car Lues at 1984 Olympics. (Economic, conservation of energy)
2. It helps in the effective organisation of the process of sport technical profession.
3. It is essential for the effective organisation of the process of development of motor abilities.
4. Need for diagonestic teaching.
5. For diagonestic coaching.

6. Necessary for the athelete themselves for self-evaluation.
7. For acceptance of equipments and change in the rules.
8. Research purposes
   (a) Evaluation of new technique.
   (b) For selection of technique in relation to their suitability to the athelete.
9. For selection of players. e.g. Short players for gymnastics and Tall players for Volley ball.
10. To understand the differences between competition exercise and training exercise as well as implement the same for best outcome.

Chapter-2

SOME BASIC MATHEMATICAL CONCEPT RELATED TO BIOMECHANICS

Algebra

It is a branch of mathematics, which deals with the signes and later to represent numbers or in place of number.

Trignometry

It is the branch of mathematics that deals with the relationship of the size and angles of a Triangle.

$$\text{Sin } \theta = \frac{\text{Oppo}}{\text{Hypo}} = \frac{b}{c}$$

$$\text{Cos} = \frac{\text{Adja}}{\text{Hypo}} = \frac{a}{c}$$

$$\text{Tan}\theta = \frac{\text{Oppo}}{\text{Adja}} = \frac{b}{a}$$

$$\text{Cot}\theta = \frac{\text{Adja}}{\text{Oppo}} = \frac{a}{b}$$

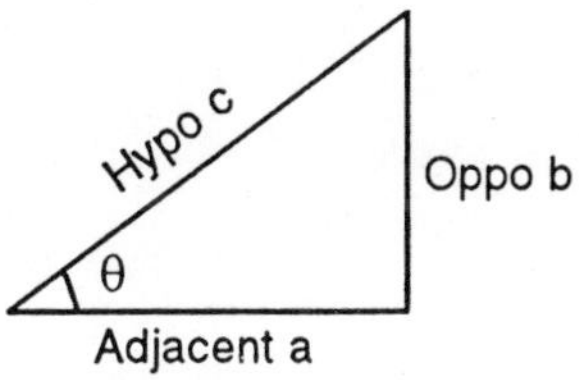

Fig. 2.1

Summation and Resolution of Forces - (Vector)

Force

Magnitudes | Direction | Point of applicatior

Summation

It is the combined effect of forces or resultant of forces- e.g. $F = F_1 + F_2$ (when force are acting at same direction)

Resulation

Breaking down the summation into individual component force.

Parallelogram Laws of Forces

If two forces are acting at a point be represented in magnitude as well as direction while a two adjacent side of a paralogram then their resultant will be represented in magnitude as well as direction in direction by the diagonal passing through that point.

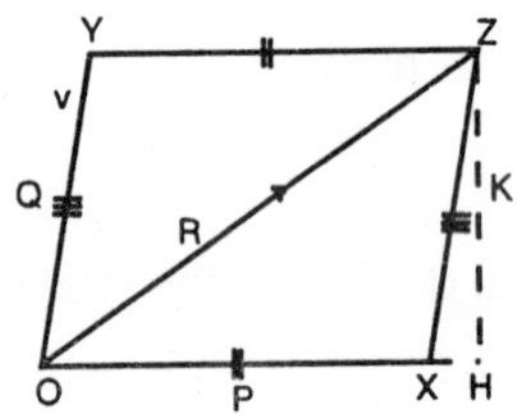

Fig. 2.3

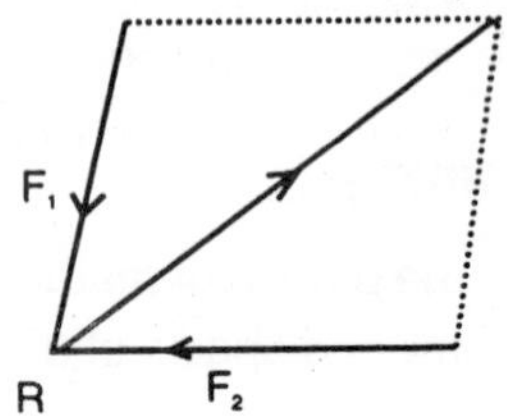

Fig. 2.2

$$R = \sqrt{P^2 + Q^2 + 2PQ \, Cos\theta}$$

$$Tan\theta = \frac{Sin\theta}{P + Cos\theta}$$

Fundamental Units

The units for any physical quantity can be diverted ultimately from the units of length, mass and time. These three units are called the fundamental units of measurements.

Fundamental units being **length, Mass** and **Time**, the units of all other physical quantity which are based on this three units are termed as **Derived Units.**

According to **M. K. S. system**

Length

The unit is metre defined as the distance between the ends of rod of platinum or Iridum preserved at the international bureau of weights and measure at SEVRES near Paris at 4°C.

Mass

The units is *Kilogram* defined as the mass of a cylinder of platinum or iridum preserved at the international bureau of weight and measure.

1 kgm = 1000 gm.

Time

The unit of time is a *second* which is based on the *mean solar day* as the standard time. Time mean solar day is divided into *24 hours.* One hour divided into 60 minutes and minute in 60 seconds.

1 mean solar day = 24 x 60 x 60 = 86,400 secs

$$\text{or } 1 \text{ sec} = \frac{1}{86,400} \text{ mean solar day}$$

Derived Units

Area = *Square metre*

Volumn = Metre^3

Speed = M/Sec.

Momentum = kgm × M/Sec.

Chapter-3

IMPORTANT VOCABULARY TO STUDY BIOMECHANICS

Acceleration : The rate of change of velocity.

Acceleration : The change in velocity (speed or direction of or both) per unit of time.

Accelerometer : A transducer that senses acceleration.

Acrodynamic Drag Force : The fluid force opposing the motion of a body moving through the air.

Acrodynamic Lift Force : The fluid force directed perpendicular to the flow direction past an object with a lift-producing shape or past a spinning ball.

Agonist : A muscle that causes a motion (mover).

Angular Velocity : The rate of rotary motion.

Anatomical Position : The upright body position in which all joints are extended and the palms are facing forward.

Angle of Attack : The angle formed between the main plane or longitudinal axis of an object and the direction of fluid flow past it.

Angle of Projection : The angle at which an object or body is projected at release or takeoff; measured in relation to a stated frame of reference, usually the horizontal.

Angular : The product of a body's rotational inertia and angular velocity.

Antagonist : A muscle that can cause the joint movement opposite to the movement being done by an agonist.

Anthropometrics: The area of study that is connected with the body's physical composition and quantifiable characteristics.

Appendicular Skeleton : A portion of the human skeleton consisting of the upper and lower extremities.

Anteroposterior (AP) Axis : An imaginary line that runs from the front to the back of the body or through an articulation.

Archimedeś Principle : A body that is partially or totally immersed in a fluid will experience an upward immersed in a fluid will experience an upward buoyant force that is equal to the weight of the volume of fluid displaced by that body.

Articulation : The junction of two bones.

Attitude Angle : The angle formed between the horizontal and the main plane or longitudinal axis of an airfoil or similar lift-producing object.

Axial Skeleton : A portion of the human skeleton consisting of the skull, thorax, and vertebral column.

Axis of Momentum : The spatial axis about which the resultant angular momentum of a rotating body exists.

Axis of Rotation : The imaginary line or point about which a body or segment rotates.

Base of Support : That region bounded by body parts in contact with some resistive surface that exerts a reaction force against the body.

Bernoulli's Principle : A low-pressure zone is created in a region of high fluid flow velocity, and a high-pressure zone is created in a region of low fluid flow velocity.

Biomechanics : The area of study between wherein knowledge and methods of mechanics are applied to the structure and function of the living human system.

Buoyant Force : The upward force exerted on an immersed body by the water beneath it.

Cardinal Axis or Plane : See Principle Axis or Plane.

Centre of Buoyancy : The point at which buoyant force acts on an immersed body's volume; the centre of gravity of the volume of water being displaced by the body.

Centre of Gravity: The point at which all the body's mass seems to be concentrated; the balance point of a body; the point around which the sum of the torques of the segmental weights is equal to zero. The point of application of gravity's force on a mass; the centre of mass.

Centre of Pressure : The point of application of a fluid lift force acting on a body moving through the fluid.

Centrifugal Force: The force directed radially outward that is exerted by a rotating body on a structure or mass that exerts a centre-directed (centripetal) force on that body.

Centripetal Force : The force directed radially toward the centre of rotation that is exerted by a mass on a rotating body and that causes that body to travel in a circular path.

Cinematography : Motion picture photography.

Closed Skill : A skill performed in a predictable environment so that the performer is free to move without having to make quick decisions.

Coefficient of Friction : The ratio of the magnitude of the maximum force of friction to the magnitude of the perpendicular force pressing the two surfaces together (normal force).

Colinear Forces : Forces whose lines of action lie along the same line.

Concentric Tension : The contraction of a muscle during which the muscle shortens and causes movement of one or more attached segments.

Constraint : A restriction to the performance of free human movement patterns; a limiting factor.

Continuous Skill : A skill in which the same movement pattern is performed repeatedly as cycles of the total act.

Centrifugal Force : The reaction force which is equal in magnitude to the centripetal force but acts in the opposite direction.

Centripetal Force : The force that deplicts a body from its linear path and compels it is move along a curve.

Circular Motion : Motion of a body along a curved path of constant radius.

Coefficient of Frictions : The ratio of the force needed to over come frictions to the normal force pressing the surfaces together.

Couple : Two forces of equal magnitude which act in opposite directions, but not in the same line.

Curvilinear : Refers to motion along a curved line or path.

Deceleration : The decrease in velocity per unit time.

Density : The mass per unit volume of an object or body (mass density); or weight per unit volume (weight density).

Digitize : To plot or identify the x and y coordinates of a point on an image for quantitative analysis, usually performed witha computerized device.

Discrete Skill : A skill in which a movement pattern is performed as a single act.

Displacement : The change in a body's location in space in a given direction.

Distal : Refers to the end of a segment, bone, or muscle attachment that is farther from the axial skeleton.

Dynamic Equilibrium : The state of a body moving with constant speed and direction, that is, with zero acceleration.

Dynamic (Isotonic) Tension : The contraction of a muscle during which the muscle changes length.

Dynamics : The study of mechanical factors associated with systems in motion.

Eccentric Tension : The contraction of a muscle during which the muscle lengthens and resists segemental motion.

Elastic : That property of a body that causes it to reform following deformation.

Electrogoniometer (ELGON) : A device that senses joint angle changes and sends electrical signals to a recording device.

Electromyography (EMG) : The technique used to sense and record levels of muscular activity through electrodes placed on the skin or inserted into the muscle.

Endurance : The ability to continue performance of a movement activity.

Energy : The ability to do work.

Equilibrant Force : The force which produces equilibrium.

Equilibrium : The state of a system whose motion is not being changed, accelerated, or decelerated.

External : Outside a defined system.

Exteroceptors : Sensory receptors located in or around the skin that respond to changes in the environment making direct contact with the exterior of the body.

Femorial Torsion	:	A relative position of the femur and tibia in which the femur is rotated medially around its longitudinal axis relative to the tibia (knees facing inward).
Flexibility	:	The range of motion permitted between two adjacent segments.
Force	:	That which produces or prevents motion, or has tendecy to do so. That which causes or tends to cause a change in a body's motion or space.
Force Arm (Moment Arm, Lever Arm, or Torque Arm)	:	The perpendicular distance between the line of action of the force and the axis of rotation.
Friction	:	A force which opposes motion. The force that resists the sliding of one surface upon another.
Frontal	:	Refers to a plane that divides a body or segment into front and back portions.
Gyroscopic Stability	:	The resistance of a rotating body to a change in its place of rotation.
Gas	:	The state of a matter which has no definite shape or volume.
Gravitional Attraction	:	The mutual force of attraction between bodies.
Impulse	:	The product of a force and the length of time acts. The product of the magnitude of a

force or torque and its time of application.

Inertia : The property of matter which requires that a force be exerted on it in order to change its position.

The resistance of a body to a change in its state of motion.

Inferior : Refers to a portion of a segment, bone, or muscle closer to the feet.

Instantaneous Velocity : The velocity of a body at a given instant in time.

Internal : Within a defined system.

Interoceptors : Sensory receptors located within the body that respond to changes in the body's internal environment and control visceral functioning.

Isokinetic : Refers to a method of strength training in which the speed of segmental rotation is kept constant throughout the range of muscular contraction.

Isokinetic Exercise : Movements performed on devices designed to mathc the motive torque applied while maintaining a constant angular speed of joint movement (also referred to as accommodating resistance exercise).

Kinanthropometrics : The area of study that is concerned with the body's physical measurements of size, shape and proportion as they relate to human movement.

Kinematics : An area of study that is connected with the time and space factors in the motion of a system.

Kinesiology : The study of human movement from an anatomical or mechanical perspective or both.

Kinesthesis : The perception of segmental and body position and movements.

Kinetic Energy : The ability of a body to do work by virtue of its motion.

Kinetic Friction : The friction that exists between two surfaces sliding past each other.

Kinetic Link Principle : The generation of high end-point velocity accomplished through the use of accelerating and decelerating adjoining links, by the use of internal and external muscle torques, applied to the segments in a sequential manner from proximal to distal, from most massive to least massive, and from most fixed to most free.

Kinetics : An area of study that is concerned with the forces that act on a system.

Lateral : Refers to the side away from the longitudinal midline of the body or segment.

Lever Arm : See Force Arm.

Lever System : A mechanism for doing work, consisting of a body with an axis of rotation and eccentrically applied forces.

Law : A statement of scientific fact concerning natural phenomena.

Linear Momentum : A property of a body in motion; the

product of a body's mass and its velocity.

Linear Motion : Motion along a line.

Motion along a straight (recti-) or curved (curvi-) line.

Liquid : A state of matter which has a definite volume, but no definite shape.

Longitudinal Axis: An imaginary line that runs along the length of a body or segment.

Mass : The measure of the quantity of matter.

The measure of a body's inertia; the amount of matter in a body.

Mass Density : Mass per unit volume.

Matter : Any thing which occupies space and has weight.

Maximal Voluntary: Contraction The maximal amount of force that is exerted by a muscle during a static contraction against an immovable resistance.

Medial : Refers to the side closer to the longitudinal midline of the body or segment.

Mediolateral (ML) Axis : An imaginary line that runs from the medial to the lateral side of the body or through an articulation.

Mobility : The ease with which an articulation, or a series of articulations, is allowed to move before being restricted by the surrounding structures.

Moment of Inertia: See Rotational Inertia.

Momentum : The product of the mass of an object and its velocity.

A system's resistance to change in its state of motion (inertia) multiplied by its velocity.

Motion : Continuing change of place or position.

Motive Force : A force that causes motion or change in shape.

Motor Unit : A single motor nerve cell and its branches, with associated muscle fibres; the smallest functional unit of muscular contraction.

Movement Pattern: A general series of anatomical movements that have common elements of spatial configuration such as segmental movements occurring in the same plane of motion.

Multiarticulate Muscle : Muscle that crosses more than two joints.

Neutral Equilibrium : The state of a body in which if the body is displaced, it will remain in the displaced position.

Normal Force : A force directed perpendicular to a surface.

Nutation : Wobbling; the tilting of a body's principal axis of rotation from its original position.

Open Skill : A skill performed in response to an unpredictable, changing environment.

Orthogonal : Refers to the orientation of lines or planes at 90 degrees to one another.

Parabolic Path : A curvilinear path (trajectory) followed by a projectile that is unaffected by air resistance.

Parallel Forces : Forces acting in the same or opposite directions.

Physics : The physical, which deals with matter which has a definite shape and volume.

Plane of Motion : The two-dimensional space cut by a moving body; the plane along which movement occurs.

Plastic : Refers to the condition of connective tissue (ligaments or tendons) when it has been stretched past its limit and will no longer return to its original shape.

Ponderal Index : A particular ratio of the human body's weight and height that is a measure of stature, or stoutness.

Potential Energy : The ability of a body to do work by virtue of its position above an object (gravitational potential energy) or by virtue of its deformation (elastic potential energy).

Power : The product of an applied force and the speed with which it is applied; the quantity of work done per unit of time.

Pressure : The ratio of force to the area over which the force is applied.

Principal (Cardinal) Axis or Plane : An axis or plane that is directed through a body's centre of gravity.

Profile Drag (Pressure Drag or Form Drag) : A net force caused by a difference in pressure from the leading side of the body to the trailing side; the greater the differential, the greater the drag.

Projection Velocity : The speed and direction of an object at the instant of projection.

Proprioceptors : Sensory receptors located in and around joints and muscles that respond to changes in position, length, tension, and acceleration of the host tissues.

Proximal : Refers to that end of a segment bone, or muscle attachment that is closer to the axial skeleton.

Radius of Gyration : A measure of the distribution of a body's or segment's mass about an axis of rotation.

Radius of Rotation : The linear distance from an axis to a point on a rotating body.

Range of Motion : The total amount of angular displacement through which two adjacent segments may move.

Reaction Force : An equal and opposite force exerted by a second body on the first in response to a force applied by the first on the second.

Recruitment : The process of increasing the number of active motor units in a muscular contraction.

Rectilinear : Refers to motion along a straight line or path.

Repetition Maximum : The maximum amount of resistance that can be moved a designated number of times.

Resistive Force : A force that resists motion or change of shape.

Resolution of Forces : The separation of a single force into two component forces acting in definite directions on the same point.

Resultant Force : The single force which has the same effect as two or more forces acting together.

Resultant Vector : The result of the composition (addition) of two or more vector quantities.

Rotary Motion : The motion of a body turning about an axis. Motion that describes a circular path about an axis.

Rotational Inertia: (Moment of Inertia) The resistance of a body to angular acceleration.

Rotatory Motion : Motion about a point which acts as pivot.

Sagittal : Refers to plane that divides a body or segment into right and left portions.

Scalar : A quantity that has magnitude only; tha is, no direction is associated with in.

Sensory Unit : A single nerve cell, with axon, branches, and receptors; such units are located within muscle, connective and cutaneous tissue.

Skill : A general movement pattern that has been adapted to the constraints of a particular activity or sport.

Skin Friction (Surface Drag) : The fluid drag force acting on the body moving through a fluid caused by viscous friction forces.

Spatial : Refers to a set of planes and axes defined in relation to three-dimensional space.

Speed : The magnitude of a body's displacement per unit of time without regard to direction.

Sprain : A tearing of ligamentous tissue.

Stability : The ability of an articulation to absorb shock and withstand motion without injury to the joint; also, the resistance to disturbance of a body's equilibrium.

Stable Equilibrium : The state of bodyh in which, if the body is displaced, it tends to return to its original position.

Static Equilibrium: The state of a body at test.

Static Friction : The frictional force generated between two objects tending to slide past each other when no motion is occurring.

Static (Isometric): Tension Muscular contraction during which no discernible segmental movement is taking place

Statics : The study of factors associated with non-moving systems.

Strain : A tearing of muscular tissue.

Strength : The ability of a muscle or group of muscles to exert force against a resistance.

Stroboscopy : A photographic technique used in the dark with a flashing light to create superimposed images on a single photograph.

Style : Individual adaptations—or modifications—to a techniques, which are unique to the individual using them.

Superior : Refers to that portion of a segment, bone, or muscle closer to the head.

Synergy : The cooperative effort of two or more muscles contracting to accomplish a single movement.

System : A body or group of bodies whose state of motion is being examined.

Tangent : A line that meets but does not intersect a curved line or surface and that is perpendicular to the radius of curvature of the arc.

Technique : A particular type, or variatioin, of the performance of the same skill.

Teleceptors : Sensory mechanisms receptive to smell, sounds, and visual stimuli.

Terminal Velocity: The constant velocity attained by a body moving through a fluid; this is due to the lack of a net force on the body.

Tibial Torsion : A relative position of the femur and the tibia in which the tibia is rotated laterally around its longitudinal axis relative to the femur (knees facing forward).

Torque : The product of a force and the length of the moment arm on which it acts.

A turning or rotatory force; the product of a force and the perpendicular distance from the line of action of the force to the axis of rotation.

Trajectory : The aerial path followed by a projectile.

Transducer : A device that transforms one kind of energy to another (e.g., a force to an electrical signal).

Transverse : Refers to a plane that divides a body or segment into top and bottom portions.

Uniarticular Muscle	:	A muscle that crosses only one joint.
Unstable Equilibrium	:	The state of a body in which, if the body in which, if the body is displaced, it tends to increase its displacement.
Variable Resistance Exercise	:	Movement performed on devices designed to vary the resistance torque applied during one excursion of a joint movement.
Vector	:	A quantity having both magnitude and direction.
Vector Composition (Addition)	:	The process of determining the net or resultant vector from two or more component vectors.
Vector Resolution	:	The process of determining component vectors from a resultant vector.
Velocity	:	The speed and direction of a body.
Videography	:	The process of capturing images on a videotape or videodisc.
Volume	:	The measure of the amount of space which matter occupies. The three-dimensional space occupied by a body (length, width, and height).
Wave Drag or Surface Drag or Skin Drag	:	A force caused by and acting against a body moving through water, forming waves at the surface of the water.
Weight	:	The measure of the attractive force of the earth for a body. The force of earth's gravitational attaction on a body's mass.
Work	:	The force applied to a body multiplied by the distance through which that force is applied.

Chapter-4

LINEAR KINEMATICS

Motion

A body is said to be in motion, if it changes its position with respect to its surroundings.

Rest

Rest, if it does not changes.

Movement

A change of place or posture.

Types of Motion

- **Linear or** Straight line Motion
  - (a) Ractilinear Motion
  - (b) Curvilinear Motion
- **Rotary or** Rotatary Motion Angular or Circular Motion
- **General Motion**

(a) Also known as straight line translation and is actually the *motion in a straight line* i.e. run 100 mtr.

(b) Also directing one point to other but following a *curved* path e.g., Long-jump (Projectile motion).

Angular or Rotatory Motion

Consist of motion or movement of a body around a centre of rotation

or

This kind of motion results when a body moves along a circular path about same line in space so that all parts of the body travel through the same angle in the same direction and in the same time. This line which may or may not pass the body itself is known as the **axis of rotation** and places at right angles to the plane of motion of the body.

e.g., circle by gymnast

Characteristics of Angular Motion

1. There should be an *Axis* of movement.
2. When the body moves all the points or parts move towards same direction with same angular distance.
3. They make equal angle at same time.

General Motion

Combination of both linear and angular motion for example, cycling, a ball rolling over ground, cylinder rolling over ground, running over treadmill etc.

Inertia

Inertia of body is the property of a body which tends to resist a change in its state of rest or motion.

In some sport higher inertia having advantage

e.g. = heavy hockey stick for full back

= Light hockey stick for forward

In 100 m. running mass (m) provides momentum (M) for better maintenance phase.

$$\because M = m \times a$$

"Mass is the quantitative measure of inertia."

Speed

The rate of change of motion or the distance covered per unit time is called speed, and is a scaler quantity – because speed gives only the magnitude but no direction

$$S = \frac{d}{t} = \text{meter per second}$$

or

The rate at which a moving object traces out its path is called its speed.

Uniform Motion

A body is said to be in uniform motion, if it covers *equal distance in equal interval of time.*

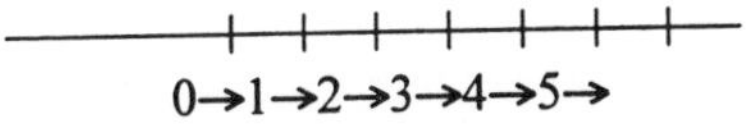

5 km. in 5 minutes, covering one kilometre per minute uniformly.

Uniform Speed

The speed of a moving object is defined to be uniform when it passes over equal lengths of its path in equal intervals of time.

Suppose that a particle describes 10 feet of its path in each second. Here the speed of the moving particle may or may not be uniform, for it is quite possible that in each second, during the first half, the particle may move over 10 feet, and during the latter half, over 5 feet, making up a total of 15 feet in each second considered. Even if it traces out 7.5 feet in each half-second, its speed may not be uniform, for there may be variations within the half-second, though the total may be the same in each case. Hence the utility of the last clause of the above definition.

When a particle is moving with uniform speed, its speed is measured by the ratio of the total length of its path traced out in any interval of time, to that time. It is the same always.

Non-uniform Speed; Speed at a Point

Consider a movinng particle whose speed is variable. For instance, an athlete starting from a station moves slowly at first and then increases its speed, and again slows it down when approaching the end line of run. We may make an estimate of the comparative quickness or slowness of its motion at two particular moments, and at one instant we may find its motion greater, than at another moment. It is quite conceivable that at a particular moment, when at a particular point of its path, it is moving with high speed, but ever after the lapse of a quarter of a second, its speed may come down considerably. Hence, it is necessary for defining *speed at any point* in such a case.

When the speed of a moving point is non-uniform, its speed at any instant during its motion, i.e., at a particular point of its path, is measured by the ratio of the distance travelled in a very short period of time, to that time, provided the time interval is infinitely small.

Let a point be moving along a path *AB*. While at any point P of its path, let it describe a small distance *PP'* (=s') in a very small time t'. Then its speed at *P* is measured by $\frac{s'}{t'}$ ultimately or more correctly by the limiting value of the ratio $\frac{s'}{t'}$ when t' is infinitely small.

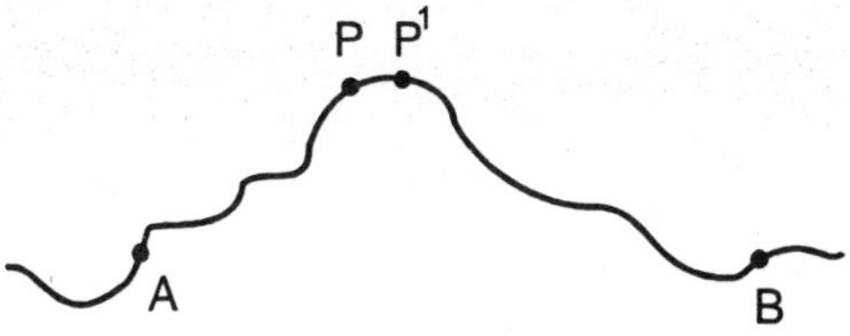

Fig. 4.1

It is clear that the speed at *P* is necessarily given by the distance actually traveller in a unit of time from *p*, for its speed may change, or it may even come to rest, before the unit time is elapsed. In fact, the speed at *P* measures the distance which the particle would pass over in a unit time, if it continued to move throughout that time at he same rate as at *P*.

Average Speed (AS)

The average speed of a moving point or an object for any prescribed interval of time during its motion, is that uniform speed with which the point or an object would describe the same length of its path as actually passed over, in the same interval of time.

Mathematically speaking, if a length AB = S of its path is described in an interval *t*, the average speed during that interval is $AS = \frac{s}{t}$.

Displacement

The displacement of a moving point in any time is its change of position, as indicated by the straight line joining its initial and final positions during the interval.

Velocity

Velocity can be defined as the rate of change of motion in a particular direction or rate of displacement is called velocity.

Meaning of Velocity

When an athelete is running in a race the rate at which he is covering his ground gives a measure of his velocity. For example he covers 12 metres in 2 seconds, then he has an average velocity or coverage rate of 6 metres per second. Thus his average velocity is defined by the equation.

$$\text{average velocity} = \frac{\text{distance covered}}{\text{time taken}} = \frac{s}{t}\ ms^{-1}$$

where *s* is the distance covered and *t* is the time taken. There is, another type of velocity, called the **instantaneous velocity.** This is the velocity that, a runner has at a particular point of time. Suppose a runner has an instantaneous velocity of 6 ms^{-1} at 2 seconds from the start of the race. This means that if his instantaneous velocity does not change he will cover 3 metres in the next second. However, suppose we find at 3 seconds it has changed and then has an instantaneous velocity of 10 ms^{-1} and has covered 9 metres during the previous second. His average velocity over this period is 9 ms^{-1} and this is not equal to either the insantaneous velocity at 2 seconds or 3 seconds. So instantaneous velocity are not necessarily equal to average velocity.

Constant Instantaneous Velocity

In fig. 4.2 (a) the instantaneous velocity at any time is the same, and must necessarily be equal to the average velocity.

$$\text{Average velocity of walker} = \frac{20}{5} = 4\ ms^{-1}$$

$$\text{Instantaneous velocity at 2 seconds say} = 2ms^{-1}$$

Note that the area under the graph is equal to the distance travelled, afact of all velocity-time graphs.

Uniformly Increasing Instantaneous Velocity

In fig. 4.2 (b) each second the instantaneous velocity increases by the same amount. When this happens the average velocity is equal to the average of the initial instantaneous velocity. (*u*) and the final instantaneous velocity (*v*). This can be shown as follows.

Distance travelled by runner = area under graph = 30 m

Average velocity of runner = $\frac{30}{4} = 7.5 \text{ ms}^{-1}$

Initial instantaneus velocity of runner = 1 m^{-1}

Final instantaneous velocity of runner = 14 ms^{-1}

$$\frac{14+1}{2} = 7.5 \text{ ms}^{-1}$$

Thus the average velocity = $\frac{u+v}{2}$ ms^{-1}

It is also equal to the instantaneous velocity at half-time as the reader can confirm: that is at 2 seconds.

Fluctuating Instantaneous Velocity

In fig 4.2 (c) the instantaneous velocity does not increase regularly and as a consequence there is no significant relationship between the average velocity of the movement and the initial and final values of the instantaneous velocity.

Average velocity of runner = $\frac{30}{5} = 6 \text{ ms}^{-1}$

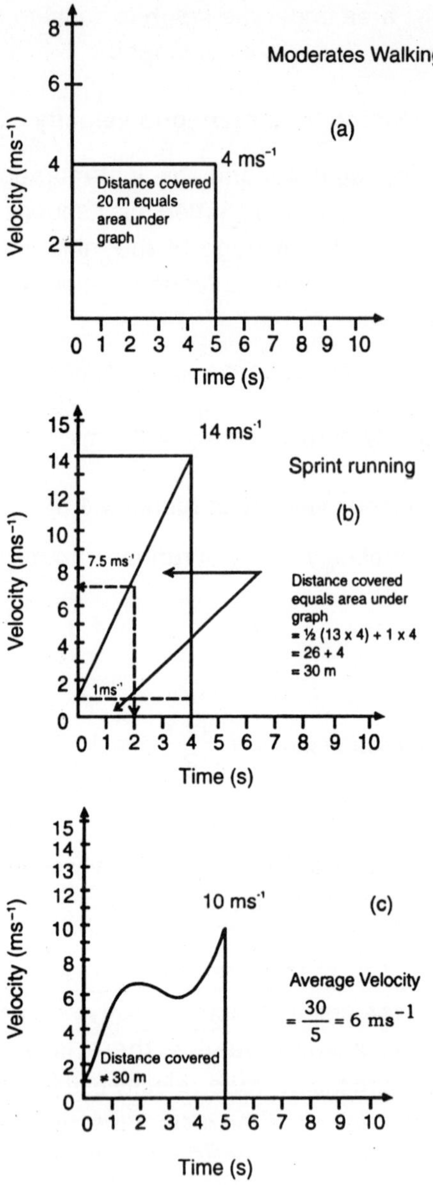

Fig. 4.2 : Velocity Time Graph

Initial instantaneous

velocity of runner = 1 ms^{-1}

Final instantaneous

velocity of runner = 10 ms^{-1}

Uniform Velocity

When an object undergoes equal displacement in each succeeding unit of time, the velocity is said to be uniform.

The velocity of a moving object is said to be uniform, when it always moves along the same straight linear or angular path in the same sense, and passes over equal linear or angular distances in equal intervals of time, however small time intervals may be taken.

Distinction between Uniform Velocity and Uniform Speed

For uniform velocity, the moving point must always move along the same direction, which need not be in the case for uniform speed, though in both cases, equal lengths of the path should be described in equal intervals of time. For instance, an object moving uniformly in a circle has uniform speed, but not uniform velocity. It may be noted that an object moving with uniform velocity must have its speed also uniform, but a point moving with uniform speed need not have a uniform velocity.

Non-uniform Velocity; Velocity at a Point

In the case when the velocity of a moving object is non-uniform, the velocity at any point p of its path [See Fig. 4.1] has go its magnitude determined by the limiting value of the ratio $\frac{s'}{t'}$ where s' the displacement PP' in an infinitely small time t' including the instant when the object is at P, and the direction of the velocity there, being the ultimate direction of PP' as P' comes infinitely close to P, is along the tangent at P to the path.

Average Velocity

The average velocity of a moving point during any interval of time is that uniform velocity with which a particle would receive the same displacement in the same time, as that of the given point.

Mathematically, if d is the length of the straight line joining the initial and final positions corresponding to the interval t, the average velocity is $\frac{d}{t}$, in the direction of that line.

It may be noted that in case of curved path the magnitude of the average velocity is different from the averge speed during any interval.

Vector and Scalar Quantities

A quantity which has got both magnitude and direction is defined to be *a vector quantity; e.g.,* velocity, displacement, force and acceleration, etc.

A quantity which has a magnitude, but is not associated with any direction, is defined to be *a scalar* quantity; e.g., speed, mass, and temperature, etc.

As a straight line has got a magnitude, direction and sense, any vector quantity may be represented by a straight line, the length of the line representing the magnitude of the vector on a suitably chosen scale, the direction of the vector, and the sense of the vector being indicated by an arrow-mark along with the line. Thus the velocity of a moving object or a force acting at a point on a body, may very well be represented in magnitude and direction by a straight line. For the sake of convenience, a velocity represented in magnitude, direction and sense by the line segment $\overline{AB}$ is often denoted by $\overrightarrow{AB}$.

Simultaneous Velocity of An Object: Resultant Velocity and Components

Let an athlete be moving on a big board of a ship. If the ship be in motion on water, the board, along with the athlete on it, shares the motion in common with the ship, and the athlete has thus two simultaneous motions. If in addition the big board dragged on the ship in another direction, the athlete also moves with the board, and it has thus three motions superposed on it. In this way we may conceive of a number of simultaneous motions of the same particle or individual.

If a particle or individual possesses several simulataneous velocities in different directions due to various reasons, and if the joint effect be the same as if the particle or individual moves with a single velocity in a definite direction, this latter velocity is known as the *resultant* of the given simultaneous velocities which, in their turn, are called the *component* velocities of the single resultant.

Parallelogram of Velocities

If an object simultaneously having two velocities which are represented in magnitude and direction by the two adjacent sides of a parallelogram meeting at an angular point, then the resultant velocity of the object is represented in magnitude and direction by the diagonal of the parallelogram drawn from that angular point.

Let a point *O* possess simultaneously two velocities *u* and *v*, represented in magnitude and direction by the two adjacent sides $\overline{OE}$ and $\overline{OF}$ of the parallelogram OEGF. Join $\overline{OG}$.

The simultaneous existence of the two velocities of the particle may be imagined by supposing that the paricle moves along $\overrightarrow{OF}$ with velocity represented by $\overline{OE}$, and at the same time, the line $\overline{OE}$, with the particle on it, moves remaining parallel to itself, so that the end 0 moves along $\overrightarrow{OE}$ with a velocity represented by $\overline{OF}$.

At the end of a unit of time, owing to the velocity u_1, the particle reaches the extremely E, while on account of the velocity v, the line $\overline{OE}$ takes up the parallel position $\overline{FG}$, so that, at the end of a unit of time, the final position of the particle is at G.

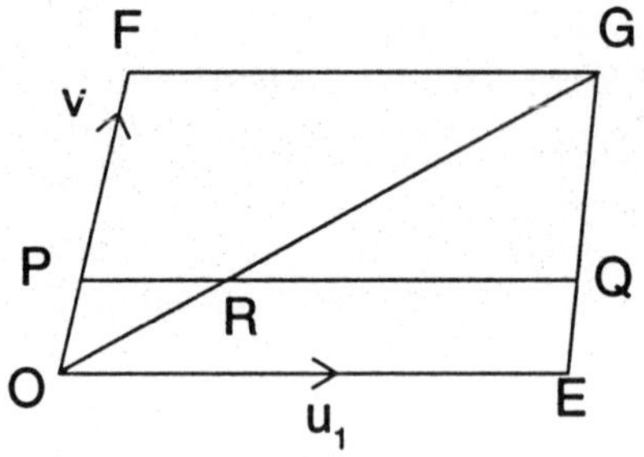

Fig. 4.3

At any intermediate instant, say after 1/*nth* of a second from start, due to the velocity v, the line $\overline{OE}$ takes up the parallel position $\overline{PQ}$, where $OP = (1/n)$. At the same time, due to the velocity u, the distance of the particle described from along PQ is $(1/n)$. OE.

Now, if R be the point of intersection of $\overline{PQ}$ and $\overline{OG}$, since $\overline{PR}$ is parallel to $\overline{FG}$,

$$\frac{PR}{FG} = \frac{OP}{OF} = \frac{1}{n}, \text{ and so } PR = \frac{1}{n}.\ OE.$$

Hence, from what is stated above, the position of the particle after 1/*n*th second from start is at *R* on $\overline{OG}$, where

$$\frac{OR}{OG}=\frac{OP}{OF}=\frac{1}{n}.$$

As *n* may have any value, we find that due to the two simultaneous velocities, *u* and *v*, the position of the particle is all along on the line $\overrightarrow{OG}$, and is such as if the particle moves with a single velocity represented by $\overline{OG}$.

$\overline{OG}$ thus represents the resultant velocity in magnitude and direction.

An Analytical Expression for the Resultant of Two Given Velocities

Let *u* and *v* be the two velocities of a particle 0 in directions

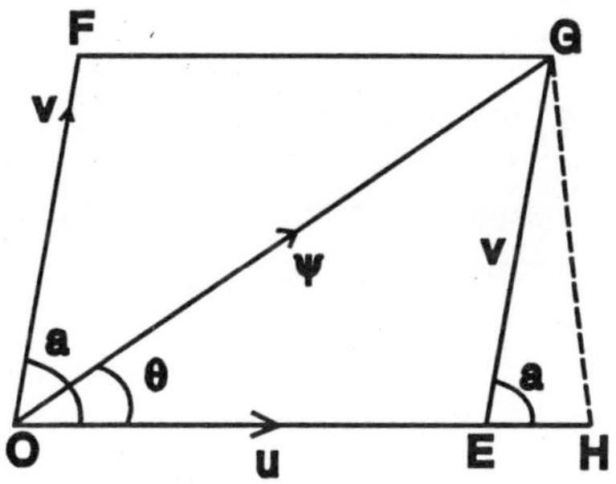

Fig. 4.4

$\overrightarrow{OE}$ and $\overrightarrow{OF}$ inclined at an angle *a*. Let them be represented by $\overline{OE}$ and $\overline{OF}$. Complete the parallelogram OEGF, and join the diagonal $\overline{OG}$, which then, by parallelogram of velocities, represents the resultant velocity Ψ. Let m $\angle GOE = \theta$ which will give the direction of resultant. Now draw $\overline{GH}$ perpendicular upon $\overline{OE}$ (produced if necessary). Then from the right-angled triangle GEH, EH = EG cos m $\angle GEH$ = v cos a, and GH = EG sin m $\angle GEH$ = v sin a.

Thus, $OG^2 = OH^2 + GH^2$ gives

$$\Psi^2 = (u + v\cos a)^2 + (v\sin a)^2$$

$$= u^2 + 2uv\,\text{Cos}\,a + v^2.$$

Also, $\tan\theta = \dfrac{GH}{OH} = \dfrac{v\sin a}{u + v\cos a}$.

Hence, $\psi = \sqrt{u^2 + 2uv\cos a + v^2}$

and $\theta = \tan^{-1}\dfrac{v\sin a}{u + v\cos a}$

give the magnitude and direction of the resultant.

cor 1. If a = 0, Ψ = u + v and if a = π, Ψ = u − v.

Hence, *the resultant of two simultaneous velocities along the same line is their algebraic sum.*

Again, if a = 90°, $\Psi = \sqrt{u^2 + v^2}$.

Cor. 2. When v = u, it is easlly seen that

$$\Psi = 2u\cos\frac{a}{2} \text{ and } \theta = \frac{a}{2}.$$

Thus, *the resultant of two equal velocities u, u at an angle a, is* $2u \cos \frac{a}{2}$ *in a direction bisecting the angle between them.*

Method of Breaking up a Given Velocity into Two Components

Any given velocity may be resolved into two components in an infinite number of ways, for parallelogram of velocities, with the straight line representing the given velocity as diagonal we construct any parallelogram, the two adjacent sides of this paralllelogram will represent the two component velocities having the given velocities as their resultant.

Further, if we want the component of a given velocity in a given direction at any inclination to it, the component is not determinable, in as much as the direction of the other component may be chosen to be anything, and the parallelogram constructed with given velocity as diagonal.

However, with a given velocity, both the directions are definitely mentioned in which required to break it up into components, and these components can be determined.

Let $\overline{OG}$ represent the given velocity V, and $\vec{OX}$ and $\vec{OY}$ given directions making angles α and β to it on either side, in which we are to resolve it into components.

Complete the parallelogram *OEGF* with diagonal $\overline{OG}$ and sides along $\vec{OX}$ and $\vec{OY}$. Then by paralleogram of velocities, $\overline{OE}$ and $\overline{OF}$ represent the required components of *V* along $\vec{OX}$ and $\vec{OY}$.

Now from triangle *OEG*, by Trigonometry,

$$\frac{OE}{\sin OGE} = \frac{EG}{\sin GOE} = \frac{OG}{\sin OEG},$$

i.e.,
$$\frac{OE}{\sin\beta}=\frac{OF}{\sin\alpha}=\frac{V}{\sin(180°-\alpha+\beta)}=\frac{V}{\sin(\alpha+\beta)}$$

$$\therefore \qquad OE=\frac{V}{\sin(\alpha+\beta)},\ OF=\frac{V\sin\alpha}{\sin(\alpha+\beta)}$$

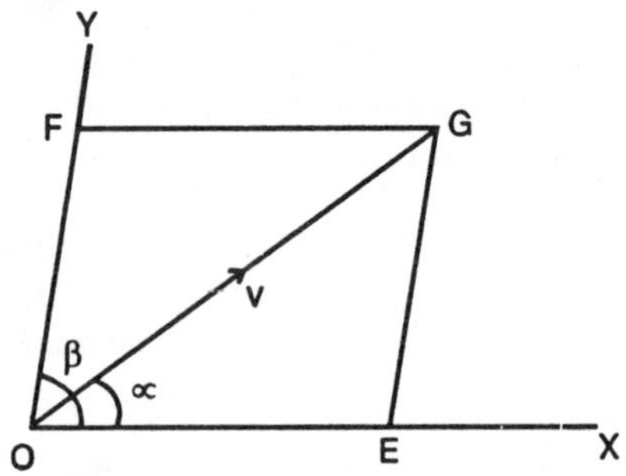

Fig. 4.5

Method of Resolving a given Velocity into Two Perpendicular Components

The most important case of resolution of a given velocity into two components when the directions of the components are at right angles to each other. In this case the components are referred to as *resolved parts* in the corresponding directions. Let the given velocity V be represented by $\overline{OG}$, and let the direction $\vec{OX}$ make an angle $GOX = \theta$ with it, $\vec{OY}$ being perpendicular to $\vec{OX}$.

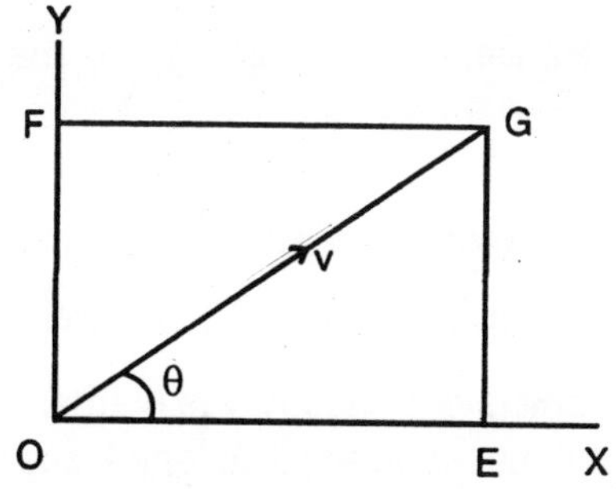

Fig. 4.6

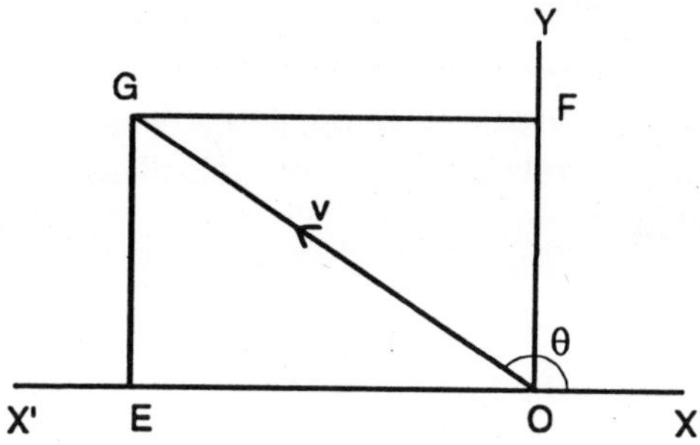

Fig. 4.7

Completing the parallelogram *OEGF* rectangle in this case, we notice that the required resolved parts along $\vec{OX}$ and $\vec{OY}$ are given by

$$OE = OG \cos m\,\angle XOG = V \cos\theta,$$

and $$OF = AG = OG \sin m\,\angle XOG = V \sin\theta,$$

Note. In figure 4.7, OE, strictly speaking, is OG cos m$\angle GOE$ = V cos (180°- θ),= -V cos θ, and is positive along $\vec{OX}$. Now, mathematically α velocity u along $\vec{OX}$ is identical with a velocity-

u along $\vec{OX}$. Hence, $-V \cos \theta'$ along $\vec{OX}$ may be described as $V \cos \theta$ along $\vec{OX}$.

So, the resolved part of V along $\vec{OX}$ is mathematically v cos θ, whether θ be obtuse or accute or anything.

Therefore, any given velocity V is mathematically equivalent to two simultaneous resolved parts, one $V \cos \theta$ along a direction $\vec{OX}$ at an angle θ to it, and another, *$V \sin \theta$ perpendicular to it,* whatever the angle θ may be. This mode of replacing a given velocity by its two equivalent resolved parts in two suitable perpendicular directions is particularly useful in finding the resultant of several simulataneous velocities for an object, as is shown below.

Determination of Resultant of Several Simultaneous Coplanar Velocities of an Object

Let an object 0 posses several simultaneous velocities $v_1, v_2, v_3,$...etc. in different given direction in the same plane and let their directions make angles θ_1, θ_2, θ_3,... with any suitable chosen direction $\vec{OX}$ in the plane, $\vec{OY}$ being perpendicular to $\vec{OX}$. (Fig. 4.8).

We can replace the velocity v_1 by its components $v_1 \cos \theta_1$ along $\vec{OX}$ and $v_1 \sin \theta_1$ along $\vec{OY}$. Similarly, v_2 may be replaced by $v_2 \cos \theta_2$ along $\vec{OX}$, and $v_2 \sin \theta_2$ along $\vec{OY}$, and so on for each one of the given velocities.

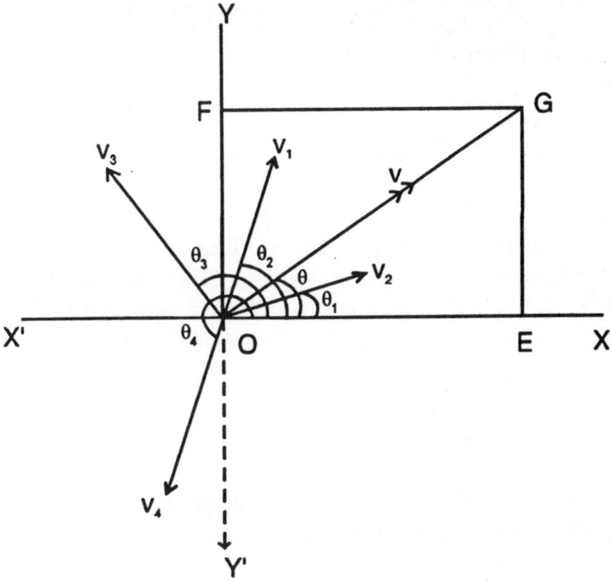

Fig. 4.8

The given simultaneous velocities are then mathematically equivalent to a single total component $v_1 \cos \theta_1 + v_2 \cos \theta_2 + v_3 \cos \theta_3 + \ldots$ along $\vec{OX}$, and a single total component $v_1 \sin \theta_1 + v_2 \sin \theta_2 + v_3 \sin \theta_3 + \ldots$ along $\vec{OY}$.

These two final components along $\vec{OX}$ and $\vec{OY}$, (represented by $\overline{OE}$ and $\overline{OF}$, say) will be equivalent to the resultant velocity V represented by $\overline{OG}$ at an angle θ to $\vec{OX}$, where

$$V \cos \theta = OE = v_1 \cos \theta_1 + v_2 \cos \theta_2 + v_3 \cos \theta_3 + \ldots$$

$$V \sin \theta = OF = v_1 \sin \theta_1 + v_2 \sin \theta_2 + v_3 \sin \theta_3 + \ldots$$

From these, V and θ are definitely determined.

Note. From the above, squaring, and adding, we ultimately get

$$V = \sqrt{\Sigma v_2^1 + 2\Sigma v_1 v_2 \cos(\theta_2 \sim \theta_1)},$$

Here, the position sign of the square root being taken for the magnitude of V, and then the signs of the right-hand expressions will give the signs of sin θ and cos θ and will thus definitely determine the direction, the direction of the resultant velocity, with the quardrant in which it lies.

Triangle of Velocities

If a moving object possess simultaneously two velocities represented in magnitude, direction and sense successively by the two sides of a triangle taken in order, their resultant is represented by the third side in opposite order.

Let $\overline{EF}$ and $\overline{FG}$ represent in magnitude, direction and sense, the two simultaneous velocities of a point. Complete the parallelogram *EFGH*. Then $\overline{EH}$, being equal and parallel to $\overline{FG}$, represents so far so as magnitude, direction and sense are concerned, the same velocity as is represented by $\overline{FG}$, Hence, the two simultaneous velocities of the moving object, being represented in magnitude and direction by $\overline{EF}$ and $\overline{EH}$, are equivalent, by parallelogram of velocities to the resultant velocity represented by $\overline{EG}$.

Note. In vector notation, $\overline{EF} + \overline{FG} + \overline{EG}$.

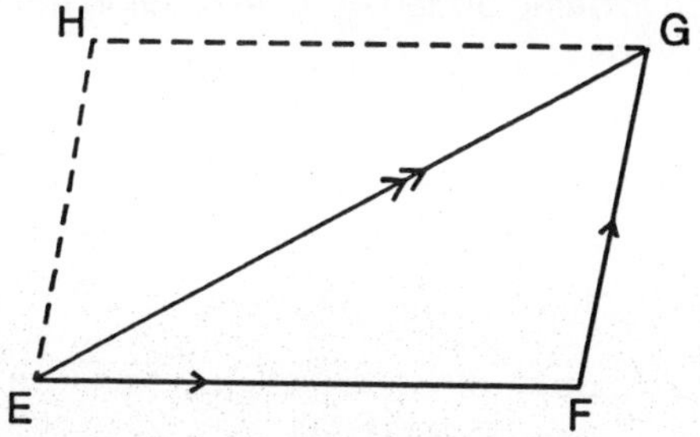

Fig. 4.9

Acceleration

The rate of change of velocity is called acceleration.May be positive. May be negative. (Negative = Declaration)

$$a = \frac{v}{t}$$

$$\text{or} = a = \frac{d}{t^2} \qquad = \frac{m}{\text{Sec}^2} \text{ uni}$$

The rate of change of velocity of a moving object is defined to be its acceleration.

Acceleratoin of a moving object is said to be uniform when equal changes of velocity in the same direction take place in equal intervals of time. First of all consider the case when an object always moves along the same straight line, but with variable velocity. For instance, consider an athlete moving along a straight track line, and suppose its velocity at a particular instant to be 10 metres per second. Three minutes later, let its velocity be observed to be 28 m^{-s}. Then in three minutes, a velocity of 18 m^{-s} is added to the original velocity in the same direction. Assuming the rate of increase to be uniform, the acceleration of the athlete is 6 m^{-s}.

It may be noted as above, that the expression for acceleration involves two units of time, one involved in the statement of the velocity which is being added and the other in the time in which it is added. The two units of time may be different, or may be same, for example, the acceleration may be described also as feet per second, or ft./sec^2. In general, in F.P.S. system, an acceleration will be expressed in ft./sec^2), and in C.G.S. system, in cms./sec^2 (or cms-sec^{-2}) units.

When the velocity of an object moving in a straight line increase, the acceleration is positive, and when it decreases, the acceleration is negative. *A negative acceleration is known as* **retardation**.

Let us consider the case when initially an object starts with a velocity in a given direction, but has a uniform acceleration in a different direction. As the velocity added in any time to the starting velocity gradually changes, and is in a direction different from that of the later, the angle made by the resultant velocity of the object with the initial direction of motion continually changes, *i.e.*, the resulting motion of the object will be along a curved path. An example of this case will be found in the motion of a projectile.

If the acceration of a moving object be *non-uniform*, it may change either in magnitude, or in directions, or in both. In case when the acceleration of a moving object is variable, the acceleration at any instant may be measured by the ultimate ratio of the change of velocity in an infinitely small time including the instant, to the time, and is in the direction in which this change of velocity takes place in the limit when the interval of time considered is infinitely small. An example of variable acceleration (where it changes in direction only) is in the case of an athlete moving in a circle with uniform speed.

When a runner is increasing or decreasing his or her velocity, he or she is said to be accelerating or decelerating respectively. The rate at which this occurs gives a value for his acceleration or decelaration. It follows for the walker of fig. 4.2 (a), that the acceleration is zero while for the runner of fig. 4.2 (b), the acceleration is constant.

The instantaneous acceleration at any particular time under these circumstances is equal to the average acceleration and both are given by the equation:

$$\text{acceleration} = \frac{\text{increase in velocity}}{\text{time taken}}$$

that is $$a = \frac{v - u}{t} ms^{-2}$$

or $$v = u + at$$

For the movement of fig. 4.2 (c) the acceleration is not constant and the above equation is not valid, except over areas where the acceleration does not change very much. Fortunately many movements can be broken down into stages where this is true and we need not take this discussion further.

Equation Governing Constant Accelaration

From previous discussion the equations below have been established to describe movement that possesses constant accelaration.

$$\text{average velocity} = \frac{s}{t}$$

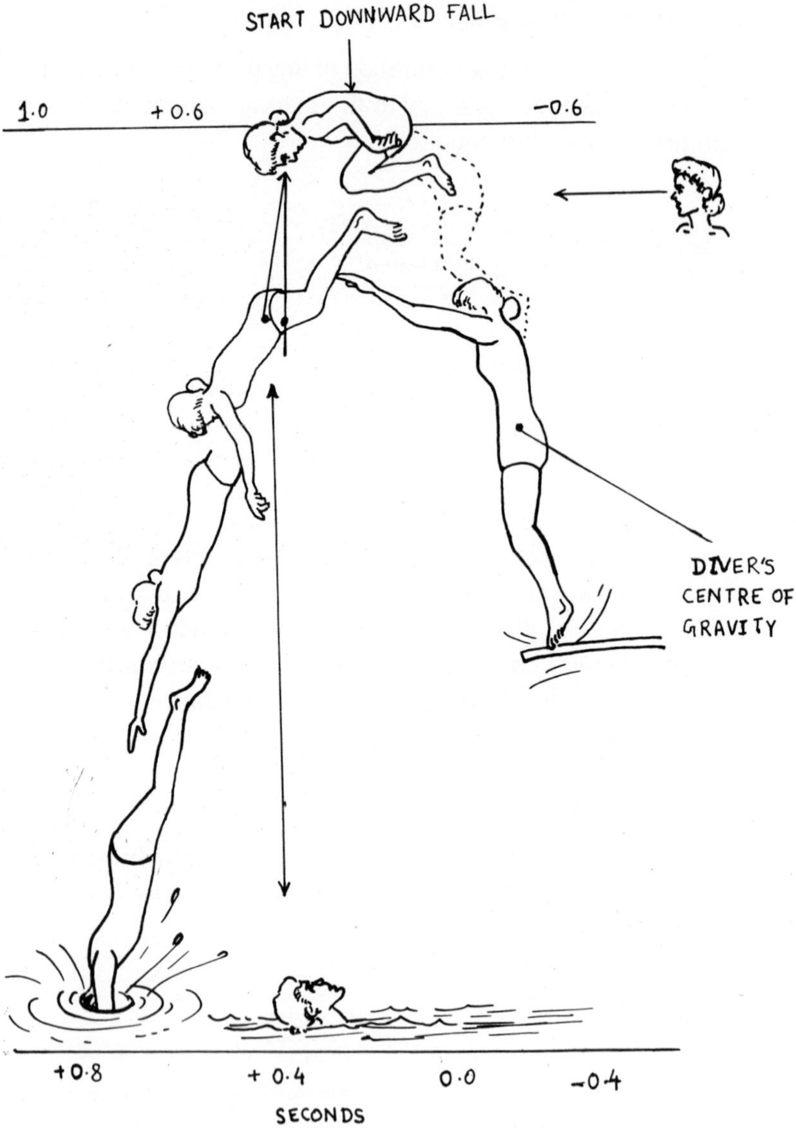

Fig. 4.10 : A diagram of a diver taken at 0.2 s. of time intervals

$$\text{average velocity} = \frac{u+v}{2}$$

$$v = u + at$$

By a contribution of these equation we get

$$s = \frac{(u+v)t}{2}$$

and $$s = vt - \tfrac{1}{2}at^2$$

$$s = ut - \tfrac{1}{2}at^2$$

and $$2as = v^2 - u^2$$

These equations cover the movement of fig. 3.2 (b). They also approximately cover verticle free fall such as that experienced in diving, jumping and trampoling. This is because the acceleration due to gravity "g" is constant and equal to 9.8 m sec^{-2}.

Consider the superimposed dive taken from a cine film shown in fig. 4.10. As the initial vertical velocity of the downward part of the dive *y* is zero and $a = 9.8$m sec^{-2}

$$s = \tfrac{1}{2}at^2$$

$$= 4.9t^2$$

Therefore for equally increasing time intervals from the start of the fall, the distance fallen, s, should increase in the ratio 1:4:9:16. It should also be noted that $v = u + at$ becomes $V = 9.8$ t and thus, where t is approximately one second, the entry velocity of the diver is about 10ms^{-1}. Which is just below a sprinter's maximum velocity as a top class sprinter can reach 13 ms^{-1} in about 2-3 seconds during a 100 m sprint.

Parallelogram of Acceleration

If a moving object possesses simultaneously two uniform acceleration represented in magnitude and direction by the two adjacent sides of a parallelogram meeting at an angular point, they are equivalent to a single resultant acceleration for moving object, which is represented in magnitude and direction by the diagonal of the parallelogram drawn from that angular point.

The proof depends on the theorem of parallelogram of velocities.

For example, let $\overline{OE}$ and $\overline{OF}$ represented in magnitude and direction the two simultaneous uniform accelerations of a moving point. Complete the parallelogram *OEGF* and join the diagonal $\overline{OG}$.

Now whatever velocity the particle might have originally, simultaneous velocities represented by $\overline{OE}$ and $\overline{OF}$ are added in each unit of time. But these two simultaneous velocities are, by parallelogram of velocities, equivalent to a single velocity represented by $\overline{OG}$. Hence, the effect is as if a single velocity $\overline{OG}$ were added in a unit time to the original velocity of the particle, and $\overline{OG}$ thus represents the resultant change of velocity of the object in an unit time. Also this rate of change of velocity must be uniform. For, at any intermediate instant, say after a time t from the initial moment, due to the two given accelerations the simultaneous changes of velocity of the particle are given by $\overline{OA}$ and $\overline{OB}$ along $\vec{OE}$ and $\vec{OF}$ respectively, where $OA = t.OE$, $OB = t.OF$. Now, if OC be taken $= t.OG$, along $\vec{OG}$ then as $\frac{OA}{OE} = \frac{OC}{OG} = \frac{OB}{OF}$, it is easily seem from geometric that *OACB* is a parallelogram, and so the two simultaneous velocities $\overline{OA}$ and $\overline{OB}$ are equivalent to a single velocity represented by $\overline{OC}$. Thus in any time t the resultant change of velocity is along $\vec{OG}$ and equal to t.OG. As this is true whatever t may be, $\overline{OG}$ represents the resultant uniform acceleration of the moving object.

Uniform Acceleration

When a body expriences the same acceleration (in both magnitude and direction) through same interval of time, its acceleration is said to be constant or uniform. e.g. freely falling bodies.

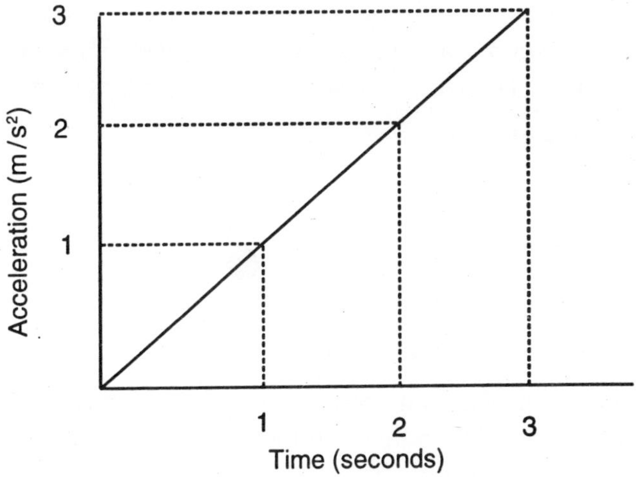

Fig. 4.12

Establish the Relationship Among, speed, Velocity, Acceleration, Force and Distance etc.

$v = u + at$ 1

$v = at$ 2

$s = ut + ½at^2$ 3

$s = \frac{1}{2}at^2$ 4

$v^2 - u^2 = 2\,as$ 5

$v^2 = 2\,as$ 6

The most important case of accelerated motion that we have to consider is, when a particle moves along the same straight line with a uniform acceleration. In this connection there are three fundamental formula which are extremely important. They are as follows:

If a point moves along a straight line with a uniform acceleration a, and if u and v denotes it s velocities at the begining and end of any interval of time t considered during its motion, and s the distance covered by it during that time, then

(i) $v = u + at$

(ii) $s = ut + \frac{1}{2}at^2$

(iii) $v^2 = u^2 + 2as$.

The proofs of these formula are given below.

I. *To prove the formula* $v = u + at$.

u being that initial and v the final velocity correspondiong to any interval of time *t* during the motion of a particle along a straight line with uniform acceleration *a* the change of velocity during the interval t is *v-u*, and this change being at a uniform rate *a*,

$$\frac{v-u}{t} = \quad ,$$

or $v = u + at.$

If the initial velocity is zero that is $u = 0$

Then, $V = at$ .... (Z)

II. *To prove the formula* $s = ut + \frac{1}{2}at^2$.

Let an object move along a straight line with uniform acceleration *a*, and let *s* be the distance described by it in any interval of time t during its motion, *u* being the velocity at the beinning, and *v*, that at the end of this interval.

As the velocity gradually changes from *u* to *v*, the average velocity during the interval is something intermediate between u and *v*. Let *V* denote its velocity at the middle of the interval, *i.e.*, at time $\frac{t}{2}$, so that

$$V = u + a\,\frac{t}{2} \qquad \dots \qquad \dots \qquad \dots \qquad \text{(i)}$$

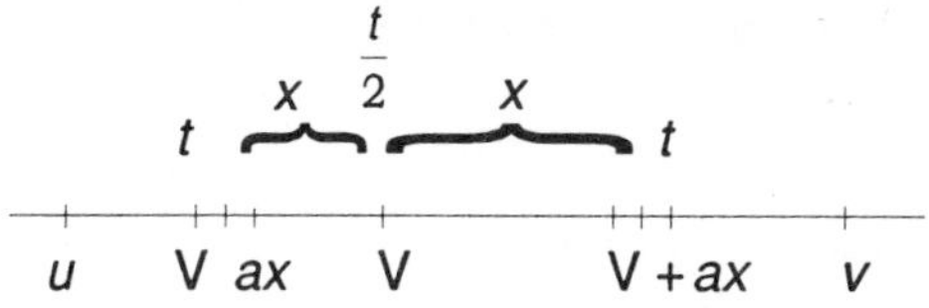

x seconds before the middle instant, the velocity is evidently $V - ax$ (a being the uniform rate at which the velocity increases), and in an extremely small interval of time τ there, the distance travelled by the object is partically $(V - ax)\tau$.

x second after the middle instant, the velocity is $V + ax$, and in an equal small interval τ here, the distance travelled is ultimately $(V + ax)x$.

Thus, the total distance describe during these two equal small intervals τ, τ is

$$(V - ax)\,\tau + (V + ax)\,\tau = 2V\tau,$$

and is the same as if the velocity were *V* during both these intervals.

As the whole time t can be divided into such pairs of equal small intervals equidistant from the middle instant, and as for each such pair the above conclusion holds, the actual distance travelled during the whole interval t is the same as if the velocity were V from beginning to end. V, therefore, represents are average velocity during the interval.

$$\text{Hence, } s = Vt = \left(u + a\frac{t}{2}\right) t = ut + \frac{1}{2}at^2 \quad \ldots\ldots \quad (3)$$

It is assumed that u equals to zero $\therefore$ u = 0

Then, $S = \frac{1}{2} at^2$ ------------ (4)

Alternative Proof

Let an object move along a straight line with uniform acceleration a, and let s be the distance described by it in any interval of time t during its motion u being the initial and v the final velocity during this interval.

Let us divide the whole time t into n equal parts, each equal to $\frac{t}{n}$.

The velocities of the object at the beginning of these successive intervals are clearly

$$u,\ u + a\frac{t}{n},\ u + \frac{2at}{n},\ \ldots\ldots\ldots\ldots\ldots\ldots,\ u + \frac{(n-1)at}{n}$$

Hence, on the assumption that the velociy of the object during each of these intervals was uniform and equal to that at the beginning of the corresponding interval, the total calculated distance that would be covered by the object is

$$s_1 = u\frac{t}{n} + \left(u + a\frac{t}{n}\right)\frac{t}{n} + \left(u + \frac{2at}{n}\right)\frac{t}{n} \ \ldots \text{ to } n \text{ terms}$$

$$= \quad u+\frac{t}{n}. \ +n+\frac{at^2}{n^2} \text{ (1 + 2 + 3 + ... to n – 1 terms)}$$

$$= ut+\frac{at^2}{n^2}\frac{n(n-1}{2}$$

$$= ut+\frac{1}{2}at^2\left(1-\frac{1}{n}\right)$$

Similarly the total distance that the object would describe as calculated on the assumption that during each of the above intervals the velocity of the object was uniform, and equal to that at the end of the corresponding interval, is

$$s_2 = \left(u+a\frac{t}{n}\right)\frac{t}{n}+\left(u+\frac{2at}{n}\right)\frac{t}{n}+ \ \ldots \text{ to } n \text{ terms}$$

$$= \frac{ut}{n}.n+\frac{at^2}{n^2} \text{ (1 + 2 + 3 + ... to n –1 terms)}$$

$$= ut+\frac{at^2}{n^2}.\frac{n(n+1}{2}$$

$$= ut+\frac{1}{2\,at^2}\left(1+\frac{1}{n}\right)$$

As actually during each interval the velocity gradually changes from its value at the beginning to that at the end, it is evident that the actual distance *s* is intermediate between s_1 and s_2, and this is true whatever value *n* may have. Making *n* infinitely large $\frac{1}{n}$ ultimately vanishes and s_1 and s_2 coincide, each being $ut+\frac{1}{2}at^2$. Now a remaining always between s_1 and s_2, must coincide with thus common value.

Hence, s = $ut + \frac{1}{2}at^2$

Proof by Graphical Method

An object travels along a straight line with uniform acceleration *a*, and let *u* be its initial velocity, *v* its velocity after time *t*, and *s* the distance travelled over during this time.

Let two mutually perpendicular straight lines $\vec{OX}$ and $\vec{OY}$ be taken as axes of reference, and let time be measured along $\vec{OX}$ and the corresponding velocity of the moving object along $\vec{OY}$.

OE = *u* along $\vec{OY}$ represents the velocity at zero-time. At any time *t*, represented by $\vec{OF}$, the velocity *v* being represented by the corresponding ordiante *PA*, we have,

$PF = v = u + at = OE + at.$

$\therefore$ $\overline{EL}$ being drawn parallel to $\overline{EL}$, $PL = at.$

$\therefore \tan \angle PEL = \frac{PL}{EL} = \frac{at}{t} = a,$ constant independent of *t*.

Hence, m $\angle PEL$ is fixed for all positions of *P*, *i.e.*, the velocity -time graph of the moving point is here a straight line $\overline{EP}$.

At any time represented by $\overline{OM}$, the velocity is given by $\overline{QM}$, and after a short interval *MM'*, it changes to *Q'M'*. If $\overline{MM}'$ be infinitely small, *Q'M'* is very nearly equal to *Q'M*, and we can take both of them ultimately equal to ½(*QM* +*Q'M'*). Hence, the distance described in time $\overline{MM}'$ is graphically represented ultimately by ½(*QM* +*Q'M'*). *M'M'*, *i.e.*, by the area of the trapezium *QM'*. Similarily, the distance travelled during the next infinite small interval *M'M'* is given graphically by the area of the trapezium *Q'M'*, and so on.

Thus the total distance travelled by the particle during the interval t (= OF), is given graphically by the area of the trapezium *OEPF*, which

= rectangle *OELF* + triangle *EPL*

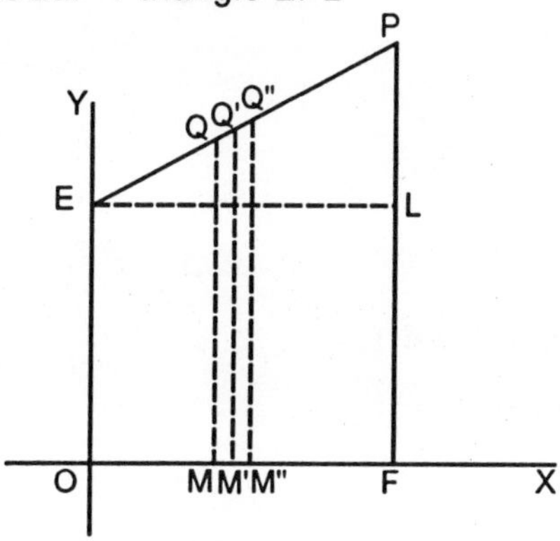

Fig. 4.13

$= OE.OF + \frac{1}{2}EL.PL$

$= ut + \frac{1}{2}t.\ at = ut + \frac{1}{2}at^2.$

Consider the average velocity of an object moving along a straight line with a uniform acceleration during any interval of time

= (*i*) *the velocity at the middle of the interval*

= (*ii*) *the mean of the initial and final velocities.*

For in this case,

$$\text{average velocity} = \frac{s}{t} = \frac{ut + \frac{1}{2}at^2}{t} = u + \frac{1}{2}at$$

$$= \frac{2u + at}{2} = \frac{u + (u + at)}{2}$$

$$= \frac{u + v}{2}$$

III. *To Prove the Formula* $v^2 = u^2 + 2as$.

Let an object be moving along a straight line with a uniform acceleration *a*, and let *u* and *v* be its velocities at the beginning and end of any interval *t* during its motion, *s* being he distance passed over during this interval.

Then we know that

$v = u + at$

and $s = ut + \frac{1}{2}af$

Hence, $v^2 = (u + at)^2 = u^2 + 2uat + a^2t^2$

$= u^2 + 2a(ut + ½at^2)$

$= u^2 + 2as$. or

$v^2 - u^2 = 2as$ .... (5)

If, it is assumed that the initial velocity *u* was zero that is $u = 0$, then it may be:

$v^2 = 2as$ .... (6)

Space Described in any Particular Second

u being the initial velocity of an object moving along a straight line with a uniform acceleration *a*, the space s_t described during the tth second may be obtained by taking the difference of the total space described in *t* seconds and the total space described in *t-1* seconds, both reckoned from the initial instant.

Thus,

$s_t = (ut + ½at^2) - \{u(t-1) + ½a(t-1)^2\}$

$= u + ½a(2t - 1)$.

Chapter-5

KINEMATICS OF FREELY FALLING BODY

OR

RECTILINEAR MOTION UNDER GRAVITY

If a heavy body is dropped from any height, it falls vertically towards the earth, its velocity, which is initially zero, continually increases as it falls, that is it falls with an acceleration. This phenomenon is attributed to the attraction of the earth on the body, known as earth's gravitiation.

If experimentations be made by dropping the body from different heights, and the corresponding times of reaching the ground be noted by an electronic stop-watch, it will be found that the distance through which the body falls is proportional to the square of the time of falling, in other words, $s = kt^2$. But this is only possible when the acceleration of the falling body is uniform, which can be proved even from fundamental considerations, hence, when a body is dropped, it falls vertically towards the earth with a constant acceleration.

If two different bodies having different mass, say a heavy stone and a light paper, be dropped from the same height, it is usually observed that the heavier body reaches the ground quicker than that of the lighter. This difference however is due to the resistance of the surrounding air which was more in the case of paper because of it's surface area. The well-known *Guinea and feather experiment of Newton,* (in which a guinea and a feather were observed by dropping them simultaneously from the same point inside a long glass tube from which air had been pumped out in advance by an air-pump) which clearly demonstrated that in the absence of air resistance, different bodies dropped from the same height reach the ground simultaneously and as each moves with a constant acceleration, it follows that at the same place on the earth *this acceleration is same for all falling bodies.*

If a body is projected upwards, it is observed that its velocity gradually diminishes, it possesses an acceleration in the opposite direction, *i.e.*, vertically downwards, the magnitude of which is the same as that of a falling body.

The above mentioned experiments, lead finally to the following conclusions:

A body free to move under the influence of earth's attraction, whether rising or falling, possesses a uniform acceleration which is vertically downwards, and this acceleration is the same vertically downwards acceleration is defined as the acceleration due to gravity, and is always denoted by 'g'. Its value may be determined by using various methods, among which the well-known pendulum experiments. It is found to vary slightly from place to place on the surface of the earth, from 32.091 ft./sec^2 at the poles. For numerical examples, this value in round figure is taken as 32 ft./sec^2 or 981 cms/sec. or 9.81 m/sec.

A Body Moving Vertically Downwards

Any problem with the help of the standard formula for uniform acceleration motion in a straight line, only replacing *a* in the formula by *g* in this case.

For example, if a body be dropped from a height *h* above the ground, the time taken to reach the ground is given by

$$h = \tfrac{1}{2}gt^2, \text{ or } t = \sqrt{\frac{2h}{g}}$$

and velocity on reaching the ground is given by

$v^2 = 2gh$, or $v = \sqrt{2gh}$

which will be referred to as the velocity due to a fall through a height h.

A Body Projected Vertically Upwards-

In this case, taking the upward direction as positive, in the formula for uniform accelerated motion in a straight line a is to be replaced by $-g$.

Here u be the velocity with which a body is projected vertically upwards. As it rises, its velocity gradually diminishes, until it becomes zero, when the body is at its greatest height. After this the body begins to fall with a gradually increasing velocity.

(i) Greatest height and the time of rise:

Let H be the greatest height attained by the body, and T the time to the greatest height.

Then, $O = u - gT$ and $O = u^2 - 2gH$.

$\therefore\ T = u/g,\ H = u^2/2g$.

(ii) Time of fall, and the velocity on reaching the ground again:

When the body begins to fall, it is at a height $\frac{u^2}{2g}$ above the ground, and its velocity is zero.

Taking the downward direction as positive, now the acceleration is + g. If .T be the time of fall, and v be the velocity on reaching the ground again, we get

$$\frac{u^2}{2g} = ½T^2 \text{ and } v^2 = 2g.\frac{u^2}{2g}.$$

Thus, $v^2 = u^2$, or $v = u$,

and $T'^2 = \frac{u^2}{g^2}$ or, $T' = \frac{u}{g} = T$.

Therefore for a body projected verically upwards:

The time of rise = The time of fall,

The velocity on reaching the ground again

= The initial velocity of projection.

Hence, accordingly *total time of flight* = $\frac{2u}{g}$

(iii) Time to a given height:

Suppose a body be projected vertically upwards with a velocity *u*, and consider *t* be the time taken to reach at a given height *h* from the starting point. Taking upward direction as positive, the acceleration of the body is $-g$.

Hence, $h = ut - \frac{1}{2}gt^2$, or $\frac{1}{2}gt^2 - ut + h = 0$.

$$\therefore \qquad t = \frac{u \pm \sqrt{u^2 - 2gh}}{g} = \frac{u}{g} \pm \frac{\sqrt{u^2 - 2gh}}{g}.$$

The reasons for this double answer is that the body is at the same height *h* twice, once on its way up, the time for which is less than *u/g*, and once on its way down, for which the time is greater than *u/g*. The difference of the two times above from *u/g* being the same, it is once more demonstrated that *the time from any point on the path to the greatest height, and the time from the greatest height back to the same point are equal.*

Further, if $h > \frac{u^2}{2g}$, the above values of *t* are imaginary, showing

that the body does not attain any height greater than that of $\frac{u^2}{2g}$,

(iv) Velocity at any height:

If a body projected vertically upwards with a velocity *u*, if *v* be the velocity at a height h above the starting point, we get, (taking the upward direction as positive),

$$V^2 = u^2 - 2gh, \text{ or, } v = \pm \sqrt{u^2 - 2gh}.$$

Distance Fallen Through in any Particular Second

Suppose a body takes t seconds to fall vertically from the highest point to the point of projection. Let S_t be the distance described in the *t* th second.

$$\therefore\ S_t = ½gt^2 - ½(t-1)^2$$

$$= ½\,(2t-1).$$

Analytical Investigation

A body is projected vertically upwards direction with a given velocity; investigate its motion.

Let u be velocity of projection.

Taking the upward directions, here the equation of motion is

$$= \frac{d^2x}{dt^2} = -g. \qquad \text{... (1)}$$

Integrating with restpect to *t*,

$$= \frac{dx}{dt} - gt + C \qquad \text{... (2)}$$

when, $t = 0$, $\frac{dx}{dt} = u$, $\therefore$ C = *u*.

$$\therefore\ \frac{dx}{dt} = u - gt. \qquad \text{... (3)}$$

∴ if *v be the velocity at time* t_i

$v = u - gt.$ ... (4)

Now integrating (3) with respect to *t*,

$x = ut - \frac{1}{2}gt^2 + D.$ ... (5)

i.e., when $t = 0$, $x = 0$, $\therefore D = 0$,

then, $x = ut - \frac{1}{2}gt^2.$ ... ... (6)

Hence, *if h be the height at time t,*

$h = ut - \frac{1}{2}gt^2.$ ... ... (7)

Eliminating *t* from (4) and (6) we get

$v^2 = u^2 - 2gx.$ ... ... (8)

Hence, if *v* be the velocity at height *h*,

then, $v^2 = u^2 - 2gh.$ ... ... (9)

From the equations (4) and (9), it is clear that as *t* and *h* gradually increase, *v* gradually diminishes and hence there comes a time and a height where this vertical upward velocity is just zero. This height is called the highest height and this time is called the time of rise. If we denote by *T* the time of rise, and by *H* the greatest height, then putting $v = 0$ in (4) and (9), then we get

$$T = \frac{u}{g} \text{ and } H = \frac{u^2}{2g} \quad ... \quad ...(10)$$

After the body reaches the greatest height, its upward vertical velocity being just zero, it is momentarily at rest and then falls freely vertically downwards direction due to the force of gravity, reaches the starting point and may even go past that point.

From (7), we have $\frac{1}{2}gt^2 - ut + h = 0$

$$\therefore t = \frac{u \pm \sqrt{u^2 - 2gh}}{g} = \frac{u}{g} \pm \frac{\sqrt{u^2 - 2gh}}{g}. \quad ...(11)$$

Now, it is clear that for a given value of h, there are two values of t *i.e.*, the body is at a given height twice, once on its way up, the time for which is less than u/g and next on its way down, for which the time is greater than u/g.

From the equation (11), it follows when $h = 0$, $t = 0$ or,

$$t = \frac{2u}{g}.$$

This reflects that the body is at the starting point twice once initially and next when it comes back to it after attaining the greatest height and the second time is called the total time of flight.

$$\therefore \text{ Total time of flight} = \frac{2u}{g}. \quad \ldots \quad \ldots (12)$$

Since time of rise $\frac{u}{g}$ and total time of flight $\frac{2u}{g}$, hence it follows that time of fall = $\frac{u}{g}$.

$$\therefore \textit{ time of rise} = \textit{time of fall} = \frac{u}{g} \quad \ldots \quad \ldots (13)$$

From the equation (9), it follows that

$$v = \pm\sqrt{u^2 - 2gh}.$$

Here, the positive sign gives that velocity on its way up and the negative sign on its way down, at *the same point of its path, the magnitude of the velocity is the same when falling or when rising.*

When, $h = 0$, $v = \pm\, u$.

This means that *the velocity on reaching the ground again = the initial velocity of projection.*

Taking the downward direction as positive, the above equation of motion (1) can be written as

$$\frac{d^2x}{dt^2} = g. \qquad \dots \qquad \dots \qquad \dots (14)$$

During the downward motion, it is usually convenient to take the highest point as origin. For example, if it is required to find time which a body takes when it is dropped freely from a height *h* above the ground.

Integrating (14) twice with respect to *t*, we have

$$\frac{dx}{dt} = gt + C \qquad \dots \qquad \dots \qquad \dots (15)$$

$$x = \tfrac{1}{2}gt^2 + Ct + D. \quad \dots \qquad \dots \qquad \dots (16)$$

Since, $\frac{dx}{dt} = 0$, $\therefore$ $C = 0$ and since $x = 0$, when $t = 0$,

$\therefore$ $D = 0$.

$$\therefore \text{ required time} = \sqrt{\frac{2h}{g}}. \qquad \dots \qquad \dots \qquad \dots (17)$$

Chapter - 6

ANGULAR KINEMATICS

Angular Distance and Angular Displacement

When a rotating body moves from one position to another, the *angular distance* through which it moves is equal to the angle between its initial and final positions, for example, a gymnast performing a forward giant circle on the horizontal bar shows the gymnast with his body in a momentary handstand position at top the bar. The sixteenth and **final** exposure shows him with about 10° yet to travel to complete the full counter-clockwise circle. Therefore, equal to 350°. The *angular displacement* that a rotating body experiences is equal in magnitude to the smaller of the two angles between the body's initial and final positions. In the example initial and final positions is 350° when measured counterclockwise and 10° when measured clockwise. The magnitude of the angular displacement is, therefore, 10°. The direction of an angular displacement can be clockwise or counterclockwise. A much smaller unit, the *degree* ($= \frac{1}{360}$ rev), is widely used in everyday life and in certain situations in sports. The *radian* (rad), is rarely used in connection with sports techniques but is widely used in engineering and other fields. If the arc *PQ* is equal in length to the radius of the circle center

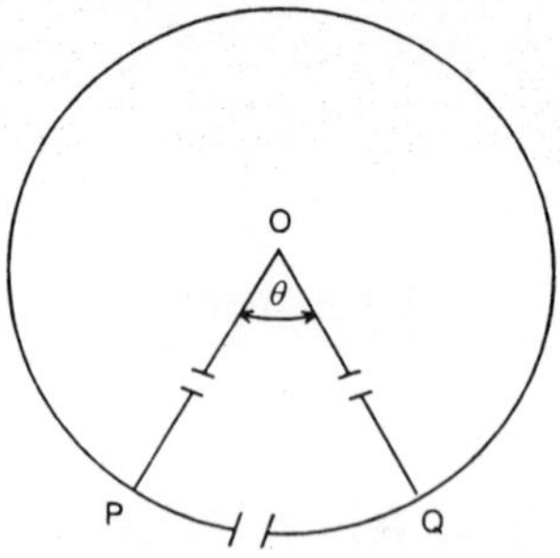

$\theta = 1$ Radian

Fig. 6.1

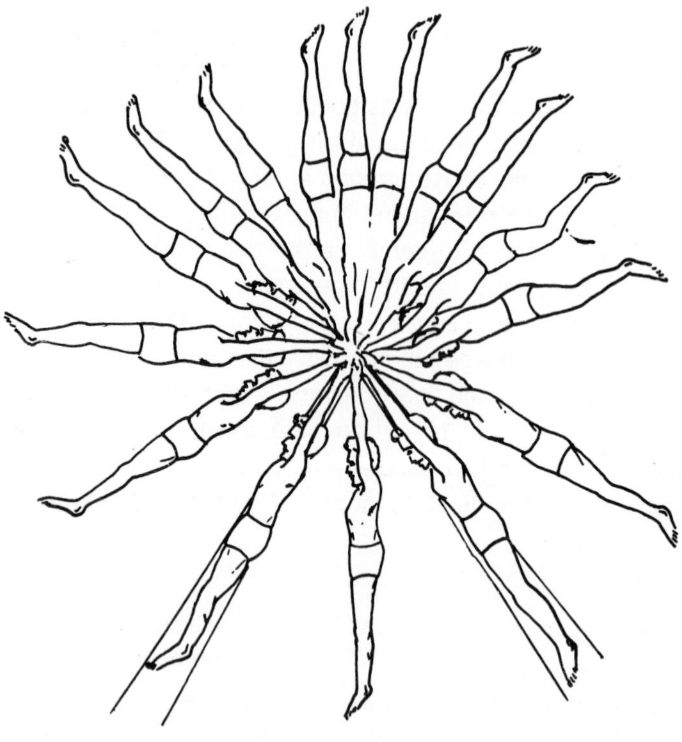

Fig. 6.2: The angular distance through which the gymnast moves

A radian is the angle subtended at the center of a circle by an arc equal in length to the radius.

Angular Velocity

Angular velocity is defined as the angular distance covered per unit time.

Angular Speed and Angular Velocity

The average angular speed of a body is obtained by dividing the angular distance through which the body moves by the time taken:

$$\bar{\sigma} = \frac{\phi}{t}$$

where $\bar{\sigma}$ = average angular speed and ϕ = angular distance.

The average angular velocity is obtained in similar fashion by dividing the angular displacement by the time taken:

$$\bar{\varpi} = \frac{\theta}{t}$$

where $\bar{\varpi}$ = average angular velocity and θ = angular displacement, and specifying the direction.

Angular Acceleration

Angular acceleration is defined as the rate of change of angular velocity. The angular acceleration, $\bar{\alpha}$, is the rate at which the angular velocity of a body changes with respect to time, In algebric form,

$$\bar{\alpha} = \frac{\varpi_f - \varpi_i}{t}$$

where ϖ_i = the initial angular velocity, and ϖ_f = the final angular velocity. Taking the gymnast as an example, if his angular velocity is 250^0/s at the instant he passes through the horizontal on his downward swing and 400^0/s as he passes below the bar 0.3s later, his average angular acceleration is

$$\bar{\alpha} = \frac{400^0 / s - 250^0 / s}{0.3s}$$
$$= 500^0 / s^2$$

Units of Angular Distance

1. R.P.M.

   Regulation per minute

2. **Degree:** One regulation is divided in 360° i.e., a degree is a part of a 360° equal parts of a regulation or circle.

3. **Radian**

   A radian is an angle which is made by an arch of a circle the length of which is equal in length to the radius of that circle.

$$\pi \text{ radian} = 180°$$

$$\therefore 1 \text{ radian} = 180 \div \frac{22}{7}$$
$$\therefore \pi = \frac{22}{7}$$
$$\text{Hence, } r = 57.29^0 \text{ (r = radius)}$$

Relationship Between Lenear and Angular Motion (Equations)

vl = vr x r .... (1)

dl = Dr x r .... (2)

Where, vl = Lenear Velocity ....m/sec

vr = Angular velocity ....red/sec

r = radius of the rotation .... Radian or metre

Dl = Leniear distance ....(metre)

Dr = Angular distance .... Radian

Linear and Angular Kinematics

This chapter began with a reference to the close relationship between the quantities used in the description of linear motion and those used in the description of angular motion.

A much more precise figure for the number of degrees in a radian can be obtained by considering how many times an arc of a length equal to the radius can be divided into the circumference of a circle:

$$\frac{\text{Circumference}}{\text{Radius}} = \frac{2\pi r}{r}$$
$$= 2\pi$$

From this it can be seen that 2π rad must be equal to 360°, or 1 rev. By simple division it can be determined that

$$1 \text{ rad} = 57.3^0$$
$$= 016 \text{ rev.}$$

Angular-Motion Vectors

A convention known as the *right-hand thumb rule* is used. In accordance with this rule, an angular-motion vector is represented by an arrow drawn so that if the curled fingers of a person's right hand point in the direction of rotation, the direction of the arrow coincides with the direction indicated by the extended thumb.

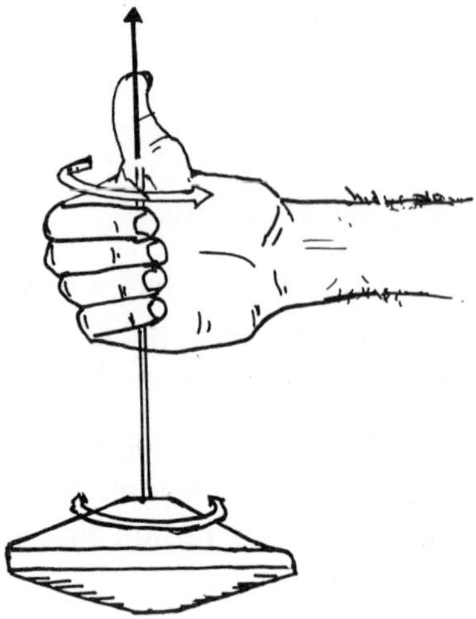

Fig. 6.3

Angular displacements cannot be summed using the parallelogram of vectors. This can readily be demonstrated by taking a book and turning it through 180° about one edge and then through another 180° about a second edge. represented by arrows and a parallelogram is constructed in the usual manner, the diagonal of that parallelogram is in no way representative of the "resultant" or total angular displacement that has taken place.

Fig. 6.4 (a):

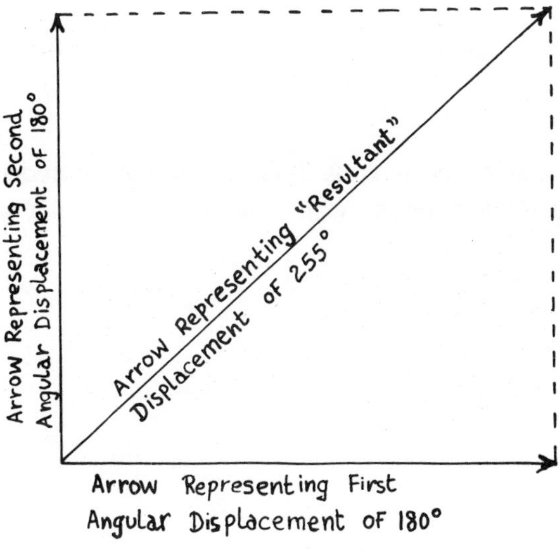

Fig. 6.4 (b): Arrow representing first angular displacement of 180°

Velocity and Angular Velocity

In many sports an athlete uses angular motion to increase the velocity of an implement. The hammer thrower whirls three or four times in the circle at an ever-increasing rate with the object of having the hammer moving at as great a velocity as possible at the moment he releases it.

In almost all projectile activities in sports the performer relies on angular motion (s) proceedings the release to obtain a specified (in the above examples, a maximum) velocity of projection. It is therefore important that the relationship between velocity and angular velocity be clearly understood.

$$\bar{s} = \frac{\text{distance}}{\text{time}} = \frac{\text{arc } PQ}{t}$$

The average angular speed of the club during the same time is given by

$$\bar{\sigma} = \frac{\text{angular distance}}{\text{time}} = \frac{(\text{arc } PQ/r)}{t} = \frac{\text{arc } PQ}{rt}$$

If the first of these equations is now substituted into the second, a relationship between the average angular speed is obtained.

Fig. 6.5: The velocity of the clubhead is equal to the product of its angular velocity and the radius of rotation

$$\bar{\sigma} = \frac{\bar{s}}{r}$$

or rearranging

$$\bar{s} = \bar{\sigma} r$$

If the distance (or time) over which the average speed and the average angular speed are computed, a similar equation relating the instantaneous speed and angular speed can be derived:

$$s = \sigma r$$

Finally, since the instantaneous speed and the instantaneous velocity of a body are equal to magnitude and the instantaneous angular speed and instantaneous angular velocity are similarly equal in magnitude the equation given above can be rewritten as follows:

$$v_y = \varpi r$$

where v_v = the velocity of the clubhead tangential to its path. This relationship between the velocity and angular velocity is of considerable importance in many sports activities. In hammer throwing, to obtain the maximum velocity of the hammer (that is, maximum v_r) by sweeping the hammer through the widest possible radium (maximum *r*), while turning as fast as can be controlled (maximum ϖ) for a constant angular velocity, the longer the radius, the greater the velocity.

Chapter - 7

NEWTON'S LAWS OF MOTION

Laws of Motion

So far we have been dealing with kinematics only, that is, we have been considering only different kinds of motions and its effects without entering into the causes which produce motions. In this chapter we will discuss the relation between the forces responsible for motion and the types of motion produced thereby. These relations are based only wholly on the three well-known laws of motion by Newton. They are as follows:

Newton's Laws of Motion

First Law (Law of Inertia)

Every body continues in its state of rest or in uniform motion in a straight line until or unless it be compelled by any external applied force to change that state.

or

A body at rest or in uniform motion will remain at rest or in uniform motion unless some external force is applied to it.

Second Law (Law of Momentum)

The rate of change of momentum is proportional to the applied force and takes place in the direction in which the force acts.

or

When body is acted upon by a constant force, its resulting acceleration is directly **proportional to the force and inversely proportional to its mass.**

or

The rate of change of momentum is directly proportional to the force applied and change takes place to the direction force applied.

$$M = \frac{mv - mu}{t} \propto F$$

$$F \propto m(v - u) \text{ or } F \propto ma$$

$$\text{or } a \propto \frac{F}{m}$$

where, a = acceleration

F = force

Ma = momentum

u = initial velocity

v = final velocity

m = mass

Third Law (Law of Reaction)

To every action, there is an equal and opposite reaction.

or

To every action there is an equal and opposite reaction force.

Action Force = Reaction Force

These laws are more or less axiomatic and have been formulated on the basis of common-sense and experience. They have been verified indirectly for bodies of ordinary size by experimenting on the results deduced from the laws.

Explanation of the First Law of Motion

This law is practically embodied in the very definition of force. A force has been defined to be that which changes (or tends to change) the state of rest or of uniform motion of a body. This means that if a body be at rest, anything which changes its state of rest is a force. Thus in absence of a force, a body at rest will continue to be at rest. Similarly, if a body is in motion, anything which changes its motion is a force, and so in absence of a force, a body in motion will have its motion unaltered, or in other words, it will continue to move uniformly.

The tendency of a body to continue in its state of rest or of uniform motion in absence of any external force. Hence, Newton's first law of motion is also referred to as the **Law of intertia**.

The first part of the law, namely that a body at rest does not move at its own unless compelled by an applied force to do so, is a matter of common experience. The second part of the law, however, that a body in motion will continue to move uniformly forever in absence of any applied force, for in the practical world we can never make a moving body absolutely and continuosly free from the influence of external forces. The tendency however of moving bodies to continue their motions when unhampered by external forces may be conceived from the following illustrations:

When a man, in a rapidly moving car, if sudden break is applied, his feet, coming in contact with the rough ground are brought to rest by the friction of the ground; but the upper part of the body, which was sharing the motion of the car, having a tendency to continue the motion, the man generally falls down.

If a galloping horse suddenly stops, the rider on its back is in danger of being thrown over the horse's head.

When a passenger sitting sideways in a car, as the car starts from rest, the part of his body in contact with the seat moves forward with the car, while the upper part of his body, having a tendency to continue in its position of rest, is generally thrown or pushed backwards. Similarly, when the car stops from motion, the upper part of his body leans or moves forward.

When a heavy piece of stones is hanging by a fine thread, the position in which the system rests is that in which the string is vertical and in this position, the forces acting on the body (namely the tension of the string and the attraction of the earth) balance one another, so that there is no resultant force left on the body. If now the stone be pulled to one side and let go, the forces on the body not balancing one another, a motion will be produced. As the system comes to the former vertical position, there will be no resultant force on the body; but in absence of a force, the body, which has already acquired some motion, does not stop and moves over to the other side. Thus, in the same position, where no resultant force area acts on the body, in one case when the body is at rest, it continues its state of rest; while in the other case, when the body is in motion, it continues its state of motion.

If a ball on a plain ground be given a motion, it is observed that after a short time, the motion of the ball comes to rest on account of the friction of the ground. Through ultimately, on account of air resistance and other fact that if it were possible to make the body free from the influence of all resisting forces, the motion once generated, would continue forever.

The first law give us a qualitative test of the existence of a force, when we find *a body moving with a uniform velocity, we concide that there is no resultant force acting on it.* On the other hand, when the motion of a body is changing either in magnitude, or in direction, there must be some force acting on it.

Explanation of The Second Law of Motion (Momentum)

Momentum

The momentum is the product of its mass and its velocity. As velocity has got a definite direction at any instant, the momentum of a moving particle has also got a magnitude and a direction, and is thus a vector quantity.

The second law of Newton aims at a quantitative measurement of a force. Now, the effect of a force on a body is generally to generate motion or to produce a change of motion. If in the same body in the same time, another force generates a greater motion, that force is greater than the former. Again, if we consider two different bodies, one heavier than the other, then it is common experience that the same force is applied to these bodies for the same time, the force required in the case of the heavier body is greater. Thus, to make an estimate of the measure of a force we are to note, not only the velocity generated by its action on a body in a definite time force also to consider the mass of the body; the measure of the force depends on the product of the mass and the velocity generated in a given time, i.e. on the momentum generated. Hence, it is the necessity of defining such a mathematical quantity as momentum. The effect of a force is to generate momentum. In case of a body of constant mass, the effect of a force is to change its velocity, i.e., to generate acceleration in it.

To deduce the formula, P = ma.

Let a force of which the measure is P be acting continuously on a body of mass m, and let v be the velocity and a the acceleration of the body at any instant during the action of the force.

Then, by Newton's second law of motion

$P \propto$ rate of change of momentum of the particle

i.e. $\propto$ rate of change of mv

i.e. $\propto m \times$ rate of change of v

[provided the mass of the body is unchanged throughout the motion].

i.e. $\propto ma$.

Hence, $P = K.\,ma$ where K is a constant. We have not as yet defined any unit for measuring a force. Let a unit of force be chosen to be that amount of force which is acting continuously on a unit mass, produces in it a unit acceleration.

Then, in the result $P = k.m.a$, when $m = 1$ and $a = 1$ by our choice, $p = 1$.

Hence, $K = 1$. Thus, when expressed in such units

$P = ma.$

Note Such a unit of force is defined to be an absolute (or dynamical unit of force).

Analytically, $P = K \dfrac{d}{dt}(mv) = Km \dfrac{dv}{dt}$

assuming m remains constant through the motion.

$\therefore\ P = k.m.a.$

Now, choosing the units suitably we have $k = 1$, as above.

$\therefore\ P = ma.$

F.P.S. and C.G.S., Absolute Units of Force

Poundal

A Poundal is that amount of force which is acting continuously on a mass of one pound, produces in it an acceleration of one foot per second. According to F.P.S. system.

Dyne

A dyne is that amount of force which is acting continuously on a mass of one gram, produces in it an acceleratoin of one centimetre per second. According to C.G.S. system.

It may be noted that the formula $P = m.a$ will be satisfied (i) when P is expressed k in poundals, m in pounds and f in ft./sec^2 or, (ii) when P is expressed in dynes, m in grammes and a in cms/sec^2. A poundal and a dyne are absolute units of force.

Relation between a poundal and a dyne:—

Since 1 foot = 30.48 cms. nearly,

and 1 lb. = 453.6 grams,

1 poundal = 30.48 x 453.6 dynes

= 13800 dynes (in round figure).

Explanation of the Third Law of Motion

Newton's Third Law of Motion gives us an insight as to how forces act in nature. It asserts that forces never exist single, but always in pair. A force may be exerted either by direct contact, as when one body presses against another or pulls it or it may be exerted as an attraction between two bodies from a distance, as in the case of earth's gravitation; or it may be of the nature of a passive resistance like friction etc. But whatever be the nature of the force, it always require two bodies or two parts of a body for the exertion of forces and corresponding to a force exerted by one part on the other, there is always an equal force exerted by the second on the first. One of these being referred to as the action, the other is called the reaction, and it must have it equal and opposite counterpart somewhere else.

One may question, "If corresponding to any force acting on a body, there is always an equal and opposite reaction, why should a body move at all?" The reason is that the action and reaction do not act on the same body. In attacking a mechanical problem, it is essential to begin by fixing upon the particular portion whose motion we are going to consider and then see which of the pair of equal and opposite actions act on this portion. Some illustrations will make this point clear.

1. When a body rests on a table, it exerts a pressure on the table on account of its weight. It itself however, experiences an upward force (or reaction) from the table which balances its weight. If the reaction were not exactly

equal to the weight, the body would move. If the table be removed, the body falls on account of its weight. Here, the action (in the form of pressure) is exerted on the table, and the reaction (in the form of the upward force of support) acts on the body.

2. Consider the typical case of a horse, dragging a carriage. As the horse pulls the carriage forwards, the carriage pulls the horse backwards, with an equal force. How is it that they ever get into motion?

The horse in attempting to draw the cart strikes the ground obliquely with its hoofs, thereby exerting a force on the ground.

The ground consequently offers an equal and opposite counterforce. Let *F*' be the horizontal component of force receieved by the horse from ground.

Fig. 7.1

Let P be the pull exerted by the horse on the carriage through the connecting striking. The horse gets an equal backward pull from the carriage. Let F' be the total resisting force on the carriage due to friction of the ground, etc. Now considering the carriage alone, it moves provided $P > F$. If m be the mass of the carriage, its acceleration a will be given by

$$P - F = m.a \quad \ldots \quad \ldots \text{(i)}$$

Considering the horse separately, it will be able to move provided $F' > P$, and if m' be the mass of the horse, the acceleration of the horse being the same as a,

$$F' - P_1 = m'a \quad \ldots \quad \ldots \text{(ii)}$$

Considering the horse and carriage together as one system P, P now from equal opposite internal forces, in the system which cancel one another, and the resultant force on the system is

$$F' - F = (m + m')\,a \quad \ldots \quad \ldots \text{(iii)}$$

Which can well be obtained from (i) and (ii). Thus, the system moves provided, only the horse can strike the ground with such a force, the forward horizontal component of the reaction received from the ground, namely F' is $> F$, the resistance to motion of the carriage. In this case, P will adjust itself to be equal to $\frac{m^1 F + mF^1}{m + m}$ as required by (i) and (ii), and f will be given by (iii).

Pressure of a Body Resting on a Moving Horizontal Plane

Let a body of mass m be placed on a horizontal plane.

Case I. When the horizontal plane is rising vertically upwards with an acceleration *a*.

The mass m on the plane rises along with the plane with same acceleration a and the force which causes it to rise is supplied due to its contact with plane in the form of the upward reaction of the plane. Assume this force of reaction to be R.

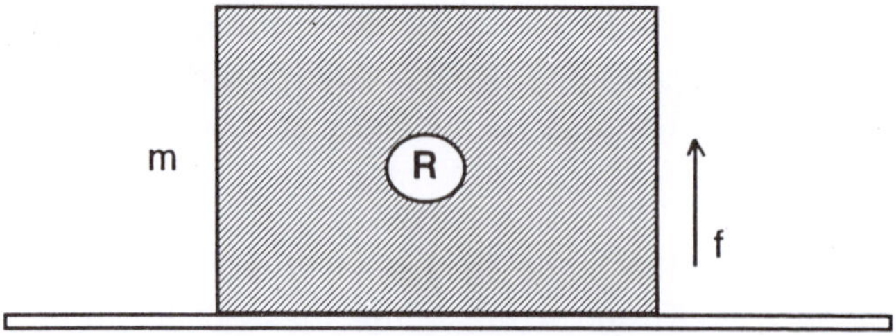

Fig. 7.2

The downward force of gravity on the body is its weight mg. Hence, the resultant upward force on it is $R - mg$ and this produces the acceleration a in the body. Hence, by Newton's second law of the motion,

$$R - mg = ma$$

or,

$$R = m(g + a)$$

Now, by Newton's third law of motion,the pressure exerted by the body on the plane is equal and opposite to the reaction of the plane on it. Hence, for the plane rising with acceleration a, the pressure exerted by a body of mass m placed on it, is

$$R = m(g + a)$$

Case II. When the horizontal plane is descending vertically downwards with an acceleration a.

The mass on the plane also descends with due same acceleration a and hence the resultant force on it is downwards so that the upward reaction on it due to its contact with due plane is less than its weight mg.

The equation for downward motion of the body separately in this case is

$$mg - R = ma$$

Hence, $R = m(g - a)$.

Analogies of Newton's Laws of Motion

or

Newton's Law of Angular Motion

Law's of Moment of Force

1. A rigid rotatory body will be continued to revolve about an axis fixed in a direction with constant angular velocity unless acted upon by some external unbalanced moment of force. e.g., Giant circle.
2. If an unbalanced moment of force acts upon a rigid bar, an angular acceleration is produced which is directly propotionate to the moment of force and inversely propotionate to the moment of intertia of the body. e.g. Rowing, i.e. reduced to create velocity.
3. For every moment of force acting on one body, there is equal and opposite moment of force about the same axis which acts upon some other body. e.g. Giant circle hard grip.

Explanation of First Law of Motion

This statement, perhaps better known as the *principle of conservation of angular momentum,* means that a body spinning will continue spinning indefinitely (and with the same angular momentum) unless some other body exerts a couple or an eccentric force on it that causes it to modify its angular motion.

This fact is of particular significant to divers, gymnasts, jumpers, and other athletes who leave the ground during the course of their events. Consider the example of a diver performing a tucked backward one-and-one-half somersault. As he leaves the board, his extended body has certain amounts of linear and angular momentum—linear momentum to project him high into the air and to give him time to complete his mid-air actions, and angular momentum to rotate him through the required one-and-one-half somersaults. Shortly after takeoff the diver moves into a tucked position, thereby moving his body mass much closer to the axis of rotation and decreasing his moment of inertia. Because his angular momentum can be altered only by an external couple or eccentric force (principle of conservation of angular momentum). A diver can decrease his moment of inertia and yet still retain the same angular

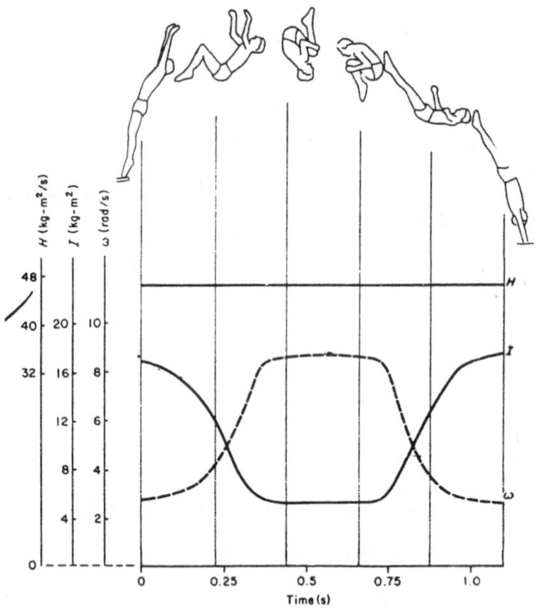

Fig. 7.3: A tucked backward one-and-one-half dive illustrates, well the interplay between angular momentum, moment of inertia, and angular velocity.

momentum is to increase his angular velocity by a corresponding amount. This is exactly what happens. At take-off he has a relatively large moment of inertia and a relatively small angular velocity. When he tucks, he decreases his moment of inertia and increases his angular velocity—and, of course, it is this latter effect that he is trying to produce. He wants to rotate quickly enough to complete his one-and-one-half somersaults in time to prepare himself for a controlled entry. As the diver comes out of his tucked position and extends his body in preparation for entry, this whole process is reversed—his angular velocity decreases as his body extension causes his moment of inertia to be increased.

Explanation of Second Law of Motion

$$T \propto \frac{I_2\omega_2 - I_1\omega_1}{t}$$

Where T = the applied torque, I_1 and I_2 = the initial and final moments of inertia, and ω_1 and ω_2 = the initial and final angular velocities. If $I_1 = I_2$, this relationship reduced to

$$T \propto I\frac{(\omega_2 - \omega_1)}{t}$$

$$\propto I\alpha$$

Where α = the angular acceleration. This equation can be converted to the form

$$T = I\alpha$$

The diver provides a useful example of the application. At the instant depicted, her body is being angularly accelerated about an axis through her feet. The torque causing this angular acceleration is equal to the product of the diver's weight and the horizontal distance x. If the diver is executing a dive requiring a large amount of rotation (say, a forward two-and-a-half), she will want this torque to be relatively large so that her body can leave the board with sufficient angular velocity to enable her to complete the dive. On the other hand, if she is executing a dive that involves very little rotation (say, a plain forward dive), she will have little need for angular acceleration during the takeoff and will want to keep the applied torque relatively small bacause her weight is constant, and the only other factor that influences the magnitude of the torque is the distance *x*, the only way in which she can control this torque is by making alterations in the magnitude of x. This is exactly what she does. By varying the amount of forward body lean that she has at take-off, she controls the magnitude of *x* and hence too the magnitude of the torque applied at this time.

Fig. 7.4: Angular action and reaction in a long-jump landing

Explanation of the Third Law of Motion

The most common is that in which an athlete applies a torque to one part of his (or her) body by contracting a muscle (or group of muscles), thereby causing that part to rotate. The equal and opposite reaction to this applied torque causes some other part of the body to rotate, or tend to rotate, in the opposite direction.

When a long jumper swings his legs forward ready for landing, a torque equal and opposite to that extend on his legs is applied to the remainder of his body. The net effect is that as the jumper swings his legs forward and upward, in say a clockwise direction, the remainder of his body moves forward and downward in a counterclockwise direction.

It should be noted here that since angular accelerations obtained when equal torques are applied to two different bodies depend on their respective moments of inertia, the effect on one body is rarely equal to that produced on the other—see the corresponding linear case. This is evident in a basketball jump shot, for example. When a player propels the ball toward the basket by extending the elbow and flexing the wrist, the remainder of the body is acted upon by torques equal in magnitude and opposite in direction to those causing these movements. Because the moment of inertia of the rest of the body is much greater than that of either the forearm or hand, the effects of these equal torques appear to be quite different. Whereas the forearm and hand sweep quickly through a relatively large angle, the remainder of the body exhibits only a slight tendency to rotate in the opposite direction.

The angular "action-reaction" effects are by no means limited to a forward-backward (or sagital) plane. In an across-the-body (or frontal) plane these effects are very clearly seen in the instinctive responses of a gymnast who senses she is about to topple off a balance beam. As soon as she feels she is starting to overbalance, she rotates her arms (and perhaps her nonsupporting leg) in the direction in which she is falling. The effect of these actions is to cause the rest of the body to rotate in the opposite direction. If this maneuver is successfully carried out, her tendency to overbalance is arrested.

Chapter – 8

PROJECTILE MOTION

Definitions

Projectile is that when any body or any object that is impelled by some force and then continues to move through the air by it's own inertia;

or

Projection of a body into the air at an angle is called the Projectile and the path of a projectile is called the *Parabole.*

In a previous chapter we have considered rectilinear motion only under gravity, as for instance when a particle is projected vertically upwards. In this chapter we shall consider free motion under gravity, when a particle is projected in any direction in space. We shall however, in order to confine overselves to a simpler case, neglect the resistance of air and consider the motion to be in vecuo. In this case, on account of the acceleration due to gravity in the vertically downward direction, the vertical component of velocity will continually change, whereas, the component of velocity in the horizontal direction will remain unchanged, as there is no acceleration in that direction. As a result, the resultant direction of motion will continually change, and the particle will describe a curved path that this path is a parabola.

A body thrown in any direction in space is defined to be a projectile, and the curved path it describes in space is called the trajectory. The initial velocity with which it is projected is defined to be the velocity of projection and the inclination to the horizon of the direction in which it is initially projected at which it falls on a plane is defined as the range on the plane, and the time interval from start till it meets the plane, *i.e.*, for which it remains in air, is called the time of flight.

Horizontal Range and Time of Flight of a Projectile

Let a particle be projected from *0* with a velocity *u*, at an angle a to the horizon. Let *R* = (*OB*) be the range on the horizontal plane through *0*, and *T* the corresponding time of flight.

The initial vertical component of velocity is clearly *u* sin a, and the acceleration in that direction on account of gravity is - *g*. After time T_1 when the particle reaches the horizontal plane through *0* at *B*, the net vertical displacement is zero, and hence confining our consideration to the motion of the particle in the verticle direction only,

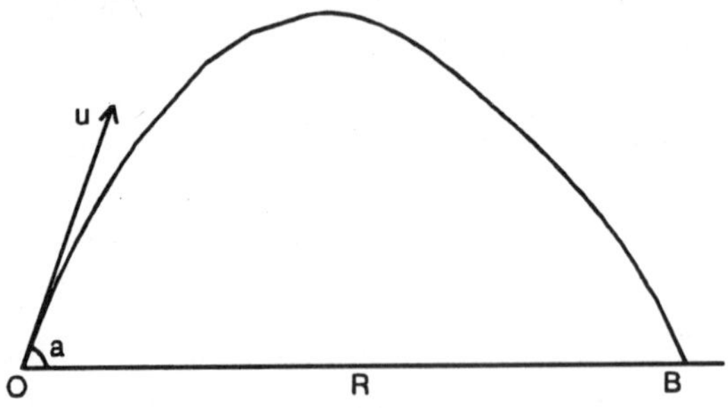

Fig. 8.1:

$u \sin \alpha, T - ½ T_2 = 0$

$$\text{giving, } T = \frac{2u \sin a}{g}$$

$$= \frac{2}{g} \text{ (initial vertical velocity)}$$

The initial horizontal component of velocity is *u* cos a, and as there is no acceleration in this direction, the above velocity remains unchanged throughout the motion. Hence,

$$OB = \frac{2u \sin a}{g}. \; u \cos\alpha = \frac{u^2}{g}\sin 2a.$$

i.e., $R = \frac{u^2}{g} \sin 2a.$

$= \frac{2}{g}$ (*initial vertical velocity x horizontal velocity*).

Cor. 1. Maximum horizontal range and direction for maximum range.

From the above value of R, it is apparent that with a given velocity of projection u, the horizontal range is greatest when sin 2a is greatest, namely unity, which requires 2a = 90°, or a = 45°.

Thus, *the maximum horizontal range* $\frac{u^2}{g}$, *is when the angle of projection is* 45°,

i.e., the direction of projection for maximum horizontal range bisects the angle between the horizontal and the vertical.

Cor.2. For a given horizontal range with a given velocity of projection, there are in general two direction of projection, equally inclined to the direction of maximum range,

Let u be the velocity of projection of a particle, and let a be the necessary angle of projection in order that the horizontal range may be a given quantity R_1.

Then $R_1 = \frac{u^2}{g}$, sin 2a,∴ sin 2a = $\frac{gR_1}{u^2}$, [a known positive quantity]

= sin θ [say, when θ is an acute angle as determined by the Trignometrical tables],

Then, as a is evidently limited to be within 90°, 2a is limited to be less than 180°, and within this limitation, 2a = θ, or, 180° θ.

∴ a = ½ a or, 90° - ½a.

Thus, there are two possible values of a, and so two directions of projection, giving the same range R_1.

Again, since the angle of projection for maximum range is 45°, and since 45° - $\frac{\theta}{2}$ = $\left(90°-\frac{\theta}{2}\right)$ - 45°, the above two directions are equally inclined to the direction of maximum range.

Note. it may be noted that if $R_1 > \frac{u^2}{g}$,sin 2a $\frac{gR_1}{u^2}$ = becomes greater than enfinity, and is impossible, *i.e.*, there is no angle of projection for range greater than $\frac{u^2}{g}$ which is really the maximum range possible.

Greatest Height Attained by a Projectile, and Time to Greatest Height

Let a particle be projected from *0* with a velocity *u* at an angle a to the horizon. Let *T'* be the time when it is at the highest point *A* of its path, and *H* (= *AM* the greates height attained).

The initial upward vertical component of velocity is u sin a, and the acceleration in this direction is - g due to gravity. The vertical velocity gradually diminishes, and at the highest point of its path A, this vertical velocity becomes zero. The corresponding time being T, and the corresponding vertical displacement being H, $0 = u \sin \alpha - gT'$ we get,

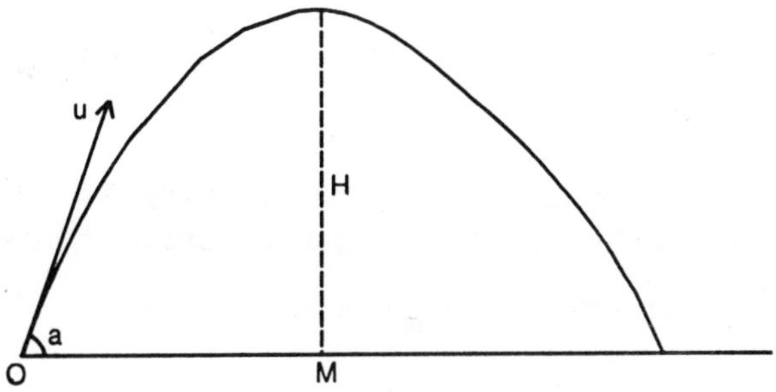

Fig. 8.2:

$$0 = \text{Sin a} - gt$$

$$\text{and } 0 = u^2 \sin^2 a - gT'$$

$$\therefore \quad T' = \frac{u \sin a}{g} \text{ and } H = \frac{u^2 \sin^2 a}{g}$$

Cor. The time to greatest height is half the total time of flight.

Position and Velocity at Any Time (t)

Let a particle be pojected from O with a velocity u at an angle a to the horizon. At any instant t after start, let P be the position of the particle, $OM(= x)$ the horizontal displacement and PM $(=y)$ the verticle displacement. Also, let v be the velocity of the projectile at P, at an angle θ to the horizon.

In the horizontal direction, the initial horizontal component of velocity is u cos a, and this remains unchanged throughout the motion,as there is no acceleration in this direction.

Hence,

$v \cos \theta$ = horizontal component of velocity at P

$= u \cos a$ ... ... (i)

and $x = OM$ = *horizontal displacement in time t*

$= u \cos a.t.$ ... ... (ii)

Again, initial vertical component of velocity is $u \sin \alpha$, and acceleration in this direction is - g due to gravity; hence,

v sin θ = vertical component of velocity at P

$= u \sin a - gt$ .... (iii)

and

y = PM = vertical displacement in time t

$= \sin a.t. - \frac{1}{2}gt^2$. ... ... (iv)

(i) and (ii) give v and θ, *i.e.*, the velocity at P in magnitude and

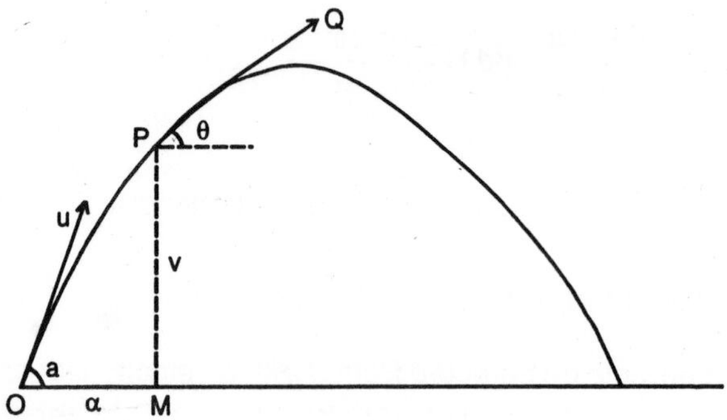

Fig. 8.3:

direction, whereas (ii) and (iv) give the position P at the instant.

The Path of a Projectile is a Parabola

First proof:

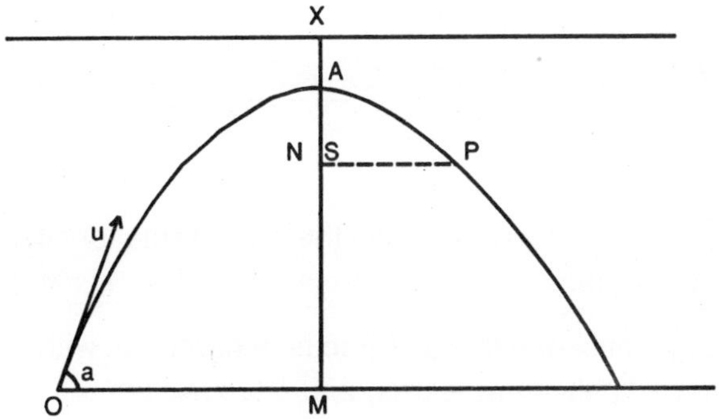

Fig. 8.4:

Let a particle be projected from *0* with a velocity *u* at an angle a with the horizon. Let *A* be the highest point of the path described by the particle, and $\overleftrightarrow{XAM}$ the vertical through *A*.

Let us measure time from the instant when the particle passes through *A*, *t* being considered positive towards the right and negative towards the left of *A*. At any instant t from *A* (positive or negative) let *P* be the position of the projectile, and *PN* the perpendicular from P on *AM*.

At *A* the vertical velocity of the particle is evidently zero, whereas, since there is no acceleration in the horizontal direction, the horizontal component of the velocity is constant throughout the motion and equal to its initial value u cos a. The acceleration vertically downwards is *g* due to gravity.

Thus,

PN = horizontal displacement of the particle in time
t = u cos a.*t*.

AN = vertical displacement from A in time t

$$= \frac{1}{2} gt^2.$$

$$\therefore \quad \frac{PN^2}{AN} = \frac{u^2 \cos^2 a.t^2}{\frac{1}{2}gt^2} = \frac{2u^2 \cos^2 a}{g},$$

a constant independent of *t*, and therefore same for all positions of P on the path.

This identifies the locus of P to be a parabola, with vertex A, axis *AM* vertically downwards, and *latus rectum*

$$4AS = \frac{2u^2 \cos^2 a}{g} = 2g \text{ (Horizontal Velocity)}^2$$

Cor. If we measure $AS = \dfrac{2u^2 \cos^2 a}{2g}$ along $\vec{AM}$, *S* is evidently the focus of the parabola. Also, measuring *AX* = *AS* above *A*, the horizontal through *X* is the directrix of the parabola.

The height of the direction above 0 = AM + AX.

$$= \frac{u^2 \sin^2 a}{2g} + \frac{u^2 \cos^2 a}{2g} = \frac{u^2}{2g}, \text{ which is independent of a.}$$

Height of the focus = MS = MA – AS

$$= \frac{u^2 \sin^2 a}{2g} + \frac{u^2 \cos^2 a}{2g} = \frac{u^2}{2g} \cos 2a.$$

Thus, *all trajectories described with same velocity of projection u from the same point in different direction in a vertical plane are parabolas with a common direction whose height above the point of projection is* $\dfrac{u^2}{2g}$, *the same as attained by a particle projected vertically upwards with the same initial velocity.*

Second Proof:

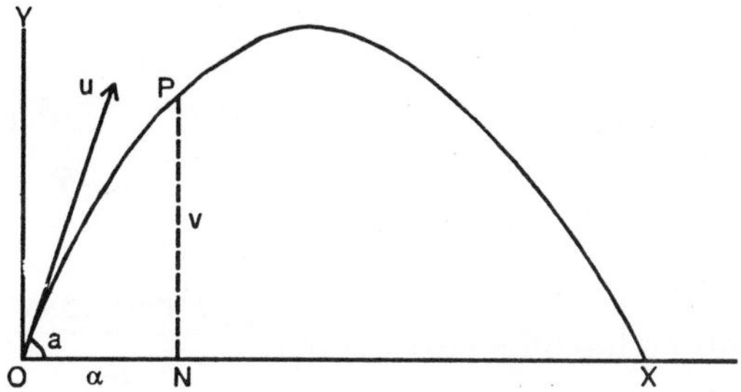

Fig. 8.5:

Let a particle be projected from *O* with a velocity *u* at an angle a to the horizon. Let the horizontal and vertical lines through *O* in the plane of projection be chosen as axes of x and y respectively. At any instant *t*, let *P* be the position of the particle whose co-ordinates are (x, y).

As there is no horizontal acceleration, the horizontal component of veloctiy of the particle is constant throughout the motion and equal to its initial value *u* cos a.

$\therefore x = u \cos a.t.$ ... ... (i)

Again, the initial vertical component of velocity is *u* sin a, and the acceleration in the vertical direction is – *g* due to gravity.

$\therefore y = u \sin a.t. - \frac{1}{2} gt^2.$ ... ... (ii)

$= x \tan a - \frac{g}{2u^2 \cos^2 a} . x2$ ... ... (iii)

Which being of the form $y = Az + Bx^2$, is th equation to a parabola. Hence, the locus of P is a parabola.

Note. The above equation of the trajectory can be written as

$$x - \frac{u^2 \sin a \cos a}{g} x = -\frac{2u^2 \cos^2 a}{g} y,$$

$$\text{i.e.} \left(x - \frac{u^2 \sin a \cos a}{g} \right)^2 = -\frac{2u^2 \cos^2 a}{g} \left(y - \frac{u^2 \sin^2 a}{2g} \right).$$

Transferring the origin to the point $\left(\frac{u^2 \sin a \cos a}{g}, \frac{u^2 \sin^2 a}{2g} \right)$

the equation becomes $x^2 = \frac{2u^2 \cos^2 a}{g} y.$

Hence, the path is a parabola whose vertex is the point $\left(\frac{u^2 \sin a \cos a}{g}, \frac{u^2 \sin^2 a}{2g} \right)$, whose latus rectum is $\frac{2u^2 \cos^2 a}{g}$ and whose axis is vertical and drawn downwards. We easily get from above

co-ordinates of the vertex: $\left(\frac{u^2 \sin 2a}{2g}, \frac{u^2 \sin^2 a}{2g} \right)$,

co-ordinates of the focus: $\left(\frac{u^2 \sin 2a}{2g}, \frac{u^2 \cos 2a}{2g} \right)$.

Alternatively, these co-ordinates may be obtained from the values of *OM, AM, SM*.

Third Proof:

Let a particle be projected from any point *B* with a velocity u in any direction $\vec{BP}$, and let $\vec{BV}$ be the verticle through *B*.

There been no gravity, the particle would move uniformly with velocity *u*, and describing a distance ut along the direction of projection would reach the point *P*. Owing to the acceleration *g* due to gravity, however, assuming the initial velocity to be absent, the particle would receive a downward vertical displacement $\frac{1}{2}gt^2$. Hence, if from *P* we take a length $PQ = \frac{1}{2}gt^2$ vertically downwards, *Q* represents the actual position of the projectile at time *t*, taking into consideration the intial velocity and the acceleration due to gravity.

Now, $\overline{QV}$ being drawn parallel to $\overline{BP}$, meeting $\overline{BV}$ at *V*,

$$QV = BP = ut$$

$$BV = PQ = \frac{1}{2}gt^2$$

$$\therefore \quad \frac{QV^2}{BV} = \frac{u^2t^2}{\frac{1}{2}gt^2} = \frac{2u^2}{g} \text{ ,a constant}$$

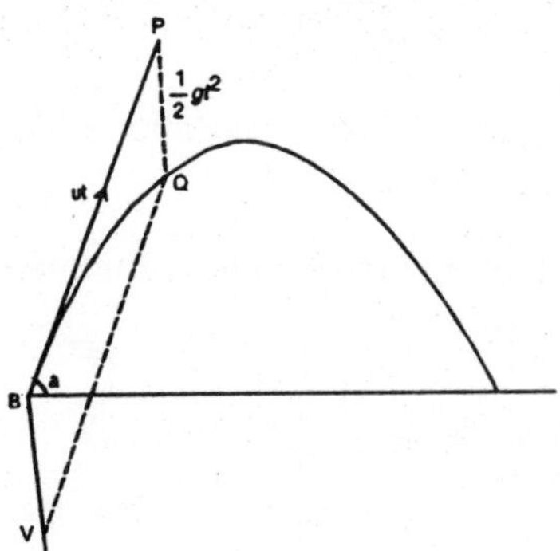

Fig. 8.6:

which, since B is fixed point, and $\overline{QV}$ are $\overline{BV}$ in fixed directions, identifies the locus of Q to be parabola.

A particle is projected horizontally from a point at any height above the ground; to show that the path described by it is a parabola.

Let a particle be projected horizontally with a velocity u from a point A, and let $\overline{AM}$ be the downward vertical through A. Let P be the position of the particle at any time t, and let PN be drawn perpendicular to AM.

Then since there is no horizontal acceleration, the horizontal velocity remains unchanged throughout the motion, namely, u. Also, at A the vertical velocity is zero,

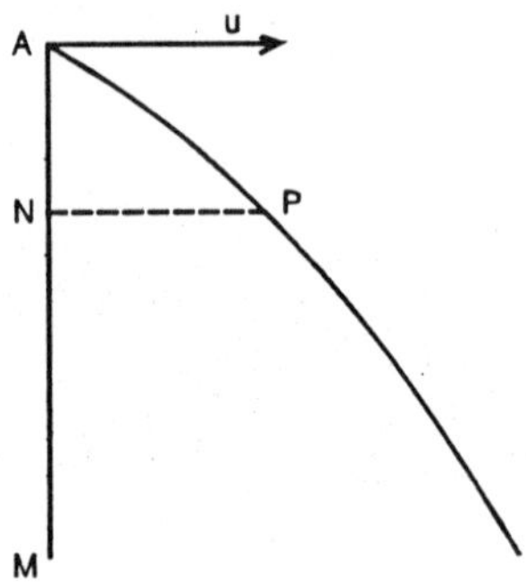

Fig. 8.7:

and due to gravity the vertical acceleration during the motion is g downwards.

Thus, PN = horizontal displacement in time t

$= ut$

and AN = vertical displacement in time t

$$= \frac{1}{2} gt^2.$$

$$\therefore \quad \frac{PN^2}{AN} = \frac{u^2t^2}{\frac{1}{2}gt^2} = \frac{2u^2}{g}, \text{ a constant,}$$

showing that the locus P is a parabola with A as vertex, $\vec{AM}$ as axis, and $\frac{2u^2}{g}$ as latus rectum.

The focus S lies on AN where $AS = \frac{1}{4}\cdot\frac{2u^2}{g}$ *i.e.* $u^2/2g$.

To find the velocity of a projectile at any point of its path at α given height from the point of projection and to show that the magnitude of the velocity is the same as that acquired by a particle allowed to fall vertically from the directrix to the point.

Let a particle be projected from 0 with a velocity u at an angle a to the horizon and let v be its velocity at an angle θ to the horizon, at a point P of its path whose height is h above 0.

As there is no horizontal accleration, the horizontal component of velocity remains unchanged throughout the motion, and so

$$v \cos\theta = u \cos a. \quad \ldots \quad \ldots \text{(i)}$$

Again, the initial upward vertical component of velocity is $u \sin a$ and $-g$ is the acceleration due to gravity in that direction. Hence, considering the motion in the vertical direction,

$$v^2 \sin^2\theta = u^2 \sin^2 a - 2gh. \quad \ldots \quad \ldots \text{(ii)}$$

From (i) and (ii), $v^2 = u^2 - 2gh$... ... (iii)

$$\text{and } \tan\theta = \pm\sqrt{\frac{u^2 \sin^2 a - 2gh}{u \cos a}} \quad \ldots \text{(iv)}$$

giving the velocity at height 'h' in magnitude and direction.

Again, since the height of the directrix of the trajectory above 0 is known to be $\frac{u^2}{2g}$, particle allowed to fall freely from the directrix to the point P describes a vertical distance $\frac{u^2}{2g}-h$, and since it moves with acceleration g due to gravity, it acquires in that time a velocty v' given by

$$v' = 2g\left(\frac{u^2}{2g}-h\right) = u^2-2gh = v^2.$$

$$\therefore\ v' = v. \qquad \ldots \qquad \ldots (v)$$

Note. Considering the magnitude of v with positive sign the double sign of tan θ corresponds to the two positions P and Q of the partical at the same height h above 0, once while rising and once while falling. Thus at these points the directions of motion make equal angles with the horizon, one above, and the other being below it.

Range on an Inclined Plane and Time of Flight.

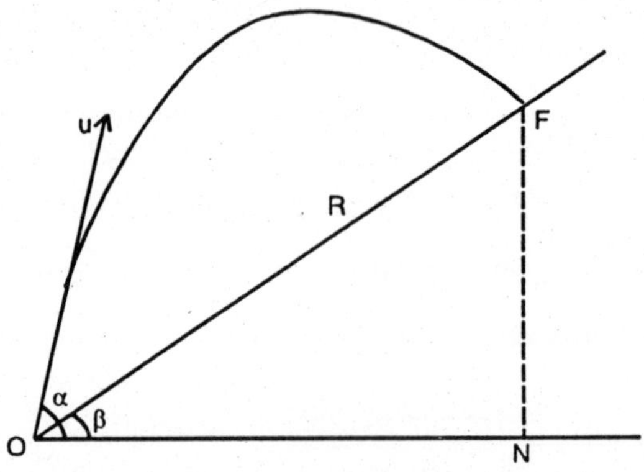

Fig. 8.8:

Let $\overrightarrow{OP}$ be an inclined plane of inclination β to the horizon through 0 from, which a partical is projected with a velocity u at an angle of protection a to the horizon, in the vertical plane through a line of greatest slope. Let the projectile, describing its path, meet the plane at *P*, so that *OP* = (= *R*) is the range on the inclined plane. Let *T* be the corresponding time of flight.

Resolving the motion along and perpendicular to the plane, the initial component of velocity perpendicular to the plane (in the upward direction) is evidently $u \sin(\alpha-\beta)$. The acceleration due to gravity is $-g$ vertically upwards, and its component perpendicular to the plane (in the upward direction) is easily seen to be $-g\cos\beta$. After time T, the particle being again on the plane, displacement of the particle perpendicular to the plane is zero, and hence,

$$0 = u\sin(\alpha-\beta).T - \tfrac{1}{2}g\cos\beta.T^2.$$

(by the formula $s = ut + \tfrac{1}{2}ft^2$)

$$\therefore T = \frac{2u\sin(\alpha-\beta)}{g\cos\beta}. \quad \ldots \quad \ldots \text{(i)}$$

As there is no horizontal acceleration, the horizontal component of velocity during this motion is constant throughout, being equal to its initial value $u\cos\alpha$, and so the horizontal displacement

$ON = u\cos\alpha.T$, *i.e.* $R\cos\beta = u\cos\alpha.T$.

$$\therefore R = \frac{u\cos\alpha}{\cos\beta}.T = \frac{2u^2\sin(\alpha-\beta)\cos\alpha}{g\cos^2\beta}$$

i.e., $$R = \frac{u^2}{g\cos^2\beta}[\sin(2\alpha-\beta)-\sin\beta]. \quad \ldots \quad \ldots \text{(ii)}$$

Alternative Method.

From the triangle *OPN*,

$$\tan\beta = \frac{PN}{ON} = \frac{\text{vertical displacement in time } T}{\text{horizontal displacement in time } T}$$

$$= \frac{u\sin\alpha . T - \frac{1}{2}gt^2}{u\cos\alpha . T} = \tan\alpha - \frac{gt}{2u\cos\alpha}.$$

$$\therefore\ T = \frac{2u\cos\alpha}{g}(\tan\alpha - \tan\beta) = \frac{2u\sin(\alpha-\beta)}{g\cos\beta} \qquad \dots (A)$$

Again, $OP = ON\sec\beta = \dfrac{u\cos\alpha . T}{\cos\beta}$

$$= \frac{u\cos\alpha}{\cos\beta}.\frac{2u\cos\alpha}{g}(\tan\alpha - \tan\beta) \qquad [\text{from } A]$$

$$\therefore \text{ range } R = \frac{2u^2\cos^2\alpha}{g\cos^2\beta}(\tan\alpha - \tan\beta)$$

$$= \frac{2u^2\cos\alpha\sin(\alpha-\beta)}{g\cos^2\beta}.$$

Cor. 1. Maximum Range on an Inclined Plane.

From the above value of *R*, with given u, the range on a given inclined plane is maximum when $\sin(2\alpha - \beta)$ is greatest, that is unity.

Hence,

$$R_{max} = \frac{u^2}{g\cos^2\beta}(1 - \sin\beta) = \frac{u^2}{g(1+\sin\beta)} \qquad \dots (iii)$$

and this occurs when $2\alpha - \beta = \frac{\pi}{2}$, or, $\alpha = \frac{1}{2}\left(\frac{\pi}{2} + \beta\right)$.

Thus, *for maximum range on an inclined plane the direction of projection bisects the angle between the horizontal and the normal to the planes, or what amounts to the same thing, the angle between the plane and the vertical.*

Cor. 2. For a given range on an inclined plane with a given velocity of projection, there are two direction of projections, equally inclined to the direction of maximum range,

From result (ii) above,

$$\sin(2\alpha - \beta) = \sin\beta + \frac{g\cos^2\beta}{u^2} R.$$

Hence, with R, u, β given, the right-hand side is known, and thus an acute angle θ of which this quantity (which is clearly positive) is the $\sin\theta$, can be determined from the Trignometrical tables. Hence,

$$2\alpha - \beta = \theta, \qquad \text{or} \qquad \pi = \theta,$$

$$\therefore \alpha = \frac{\theta}{2} + \frac{\beta}{2}, \qquad \text{or,} \qquad \frac{\pi}{2} + \frac{\theta}{2} + \frac{\beta}{2},$$

giving two possible values of α (within the range 0 to π), i.e., two possible directions of projection.

Again, the angle of projection for the maximum range on the plane is $\frac{\pi}{4} + \frac{\beta}{2}$, and since

$$\left(\frac{\pi}{4} + \frac{\beta}{2}\right) - \left(\frac{\theta}{2} + \frac{\beta}{2}\right) = \left(\frac{\pi}{2} - \frac{\theta}{2} + \frac{\beta}{2}\right) - \left(\frac{\pi}{4} + \frac{\beta}{2}\right) = \frac{\pi}{4} - \frac{\theta}{2},$$

the latter part of the result follows.

Range Down An Inclined Plane and the Time of Flight

Suppose a particle is projected from the point *P* with velocity *u*, at an angle α to the horizontal $\vec{PH}$.

Then its velocity along $\vec{PO}$ (*i.e.*, down the plane) is $u\cos(\alpha+\beta)$ and an acceleration g sin β along $\vec{PO}$ and its velocity perpendicular to $\vec{PO}$ in the upward direction is u sin $(\alpha+\beta)$ and an acceleration $g\cos\beta$ perpendicular to $\vec{PO}$ in the downward direction. Let *T'* be the time (*down* the plane), *i.e.*, from *P* to *0*. Since after time *T'*, the particle is again on the plane, the displacement of the particle, perpendicular to the plane is zero.

$$\therefore\ 0 = u\sin(\alpha+\beta)\,T' - \frac{1}{2}g\cos\beta . T'^2$$

$$\therefore\ T' = \frac{2u\sin(\alpha+\beta)}{g\cos\beta}. \qquad \ldots \qquad \ldots (1)$$

Similarly, it can be shown that if *R'* be the range down the plane *(PO)*, then

$$R' = \frac{2u^2\cos\alpha\sin(\alpha+\beta)}{g\cos^2\beta}.$$

$$= \frac{u^2}{g\cos^2\beta}[\sin(2\alpha+\beta)+\sin\beta]. \qquad \ldots (2)$$

Similarly, we obtain the maximum value of *R'*,

$$R'_{max} = \frac{u^2}{g(1-\sin\beta)}. \qquad \ldots (3)$$

Concept Based on Projectile

1. Horizontal distance (length) will be more, when

   (a) *i.e.*, depending on angle of projectile (which should be close to 45°.

   (b) Flatness or Acquitness of angle of projection will result to less or — more distance (horizontal range) respectively.

   (c) If velocity of projection is greater then there will be more distance covered.

(2) If height is aim then the angle of projection should be very close vertical.

Important Formulas

$$R = \frac{v^2 \sin 2\theta}{g}$$

Where, R = Range

V = Initial velocity

θ = Angle of projection

g = Acceleration due to gravity

$$Tf = \frac{2v \sin 2\theta}{g}$$

Where,

Tf = Time of flight,

$$h = \frac{v^2 \sin 2\theta}{2g}$$

Where,

h = Maximum higher grained by projectile,

Principles Based upon Projectile

1. Where horizontal distance is the arm the maximum velocity should be achieved at the time of release and the angle of projectile should be 45° because 2 sinθ = 90° = 1.
2. Where height is aim the angle of projectile should be close to verticle that is 90°.

Relating Projectile Motion with Freely Falling Body Kinematics

When a body is projected upwards, it is obseved that its velocity gradually diminished, or in other words, it possesses an acceleration in the opposite direction, *i.e.,* vertically downwards, the magnitude of which is the same as that of a falling body.

The above experiments, as also more careful and accurate experiments of modern times lead finally to the following conclusions:

A body free to move under the influence of earth's attraction, whether rising or falling, possesses a uniform acceleration which is vertically downwards, and this acceleration is the same for all bodies at the same place on the surface of the earth. This vertically downwards acceleration is defined as the **acceleration due to gravity,** and is always denoted by **'g'.** Its value has been determined accurately by various methods, among which mention may be made of the well-known pendulum experiments. It is found to vary slightly from place to place on the surface of the earth, from 32.091 ft./sec^2 at the equatro to 32.252 ft./sec.^{2} at the poles.* For numerical examples, this value in round figure is taken as **32 ft./sec.^{2} or 981 cms/sec.**

A Body (Projected) Moving Vertically Downwards

Taking the downward direction as positive, any problem in this case can be worked out with the help of the standard formule for uniformly accelerated motion in a straight line, only replacing *f* in the formula by *g* in this case.

For instance, if a body be dropped from a height *h* above the ground, the time taken to reach the ground is given by

$$h=\frac{1}{2}gt^2, \text{or } t=\sqrt{\frac{2h}{g}}$$

and velocity on reaching the ground is given by

$$v^2 = 2gh, \text{or } v = \sqrt{2gh}$$

which will be referred to as the velocity due to a fall through a height *h*.

A Body Projected Vertically Upwards

In this case, taking the upward direction as positive, in the formula for uniformly accelerated motion in a straight line, *f* is to be replaced by $-g$.

Let *u* be the velocity with which a body is projected vertically upwards. As it rises, its velocity gradually diminishes, untill it becomes zero, when the body is at its greatest height. After this the body begins to fall with a gradually increasing velocity.

(i) Greatest height and the time of rise :

Let *H* be the greatest height attained by the particle, and *T* the time to the greatest height.

Then, $0 = u - gT$ and $0 = u^2 - 2gH$.

$\therefore T = u/g, \qquad H = u^2/2g.$

(ii) Time of fall, and the velocity on reaching the ground again:

When the particle begins to fall, it is at the height $\frac{u^2}{2g}$ above the ground, and its velocity is zero.

Taking the downward direction as positive, now the acceleration is $+g$. If T' be the time of fall, and v be the velocity on reaching the ground again, we get

$$\frac{u^2}{2g}=\frac{1}{2}T'^2 \qquad \text{and} \qquad v^2=2.g.\frac{u^2}{2g}$$

Thus, $v^2 = u^2$, or, $v = u$,

and $T'^2=\frac{u^2}{g^2}$, or $T'^2=\frac{u}{g}=T.$

Thus, for a body projected vertically upwards,

the time of rise = the time of fall,

the velocity on reaching the ground again

= the initial velocity of projection.

Note. Accordingly, *total time of flight* $=\frac{2u}{g}$.

(iii) Time to a given height :

Let a body be projected vertically upwards with a velocity u, and let t be the time at which it is at a given height h from the starting point. Taking upward direction as positive, the acceleration of the body is $-g$.

Hence, $h = ut - \frac{1}{2}gt^2$, or, $\frac{1}{2}gt^2 - ut + h = 0.$

$$\therefore \quad t = \frac{u \pm \sqrt{u^2 - 2gh}}{g} = \frac{u}{g} \pm \frac{\sqrt{u^2 - 2gh}}{g}.$$

Note. The reasons for this double answer is that the body is at the same height h twice, once on its way up, the time for which is less than u/g, and once on its way down, for which the time is greater than u/g. The difference of the two times above from u/g being the same, it is once more demonstrated that *the time from any point on the path to the greatest height, and the time from the greatest height back to the same point are equal.*

Again, if $h > \frac{u^2}{2g}$ the above values of t are imaginary, showing that the particle does not attain any height greater than $\frac{u^2}{2g}$.

(iv) Velocity at any height :

In case of a body projected vertically upwards with a velocity u, if v be the velocity at a height h above the starting point, we get, taking the upward direction as positive,

$$v^2 = u^2 - 2gh, \quad \text{or,} \quad v = \pm\sqrt{u^2 - 2gh}.$$

Distance Fallen Through in Any Particular Second

Suppose a particle takes t seconds to fall vertically from that highest point of projection. Let S_t be the distance described in the tth second.

$$\therefore St = \frac{1}{2}gt^2 - \frac{1}{2}g(t-1)^2$$
$$= \frac{1}{2}g(2t-1).$$

Chapter - 9

KINETIC

9.1: EQUILIBRIUM

or Balance

or Stability

Definition

A body is said to be in equlibrium if the resultant of the forces acting on it is equal to zero.

Categories of Stability

Static
The equlibrium at rest is called static equlibrium.

Dynamic
The equilibrium in dynamic condition is called dynamic equilibrium.

Types of Equlibrium

a) Stable Equilibrium e.g. C.G. Ht.

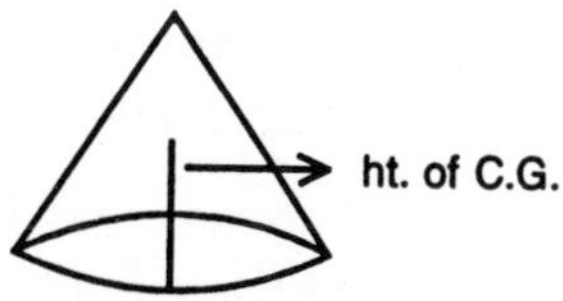

b) Unstable Equilibrium e.g., Base

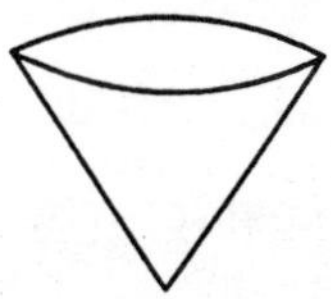

c) Neutral Equilibrium e.g. Horizontal 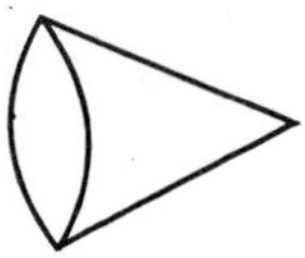

Conditions of Static Equilibrium

1. Stability is directly proportionate to the area of the base on which the body rest. e.g., bigger base - Greater Stability
2. Stability in a given direction is directly proportionate to the horizontal distance of the C.G. (Centre of gravity) from the edge of the base towards the given direction of movement.
3. Stability is directly proportionate to the weight of the body.

   e.g., Heavy object - more stability.
4. C.G. should be within the area of base.
5. Stability is indirectly proportional to the height of the C.G. from the base.

Conditions of Dynamic Equilibrium

1. During running type of movements - quick steping is necessary to protect a forward fall.
2. Opposite limb movements — also counter react the loss of equilibrium during directional changes.
3. Inward lean is required to maintain equilibrium during rapid directional changes.
4. To maintain dynamic equilibrium, the required amount of friction from the base is essential and it is related to the cofficient of friction. e.g., 'c'.

   more 'c' - difficult to move

   Less 'c' - chances of slipping

5. The dynamic equilibrium depends to a greater extent on the physiological factors - concern with the maintenance of posture and equilibrium especially.

Principle Derived from the Concept of Equilibrium

1. To start quickly in one direction keep C.G.as high as possible as near (close) to the edge of the base in the direction of movement. e.g., Crude start i.e., start posture for sprint.
2. For greater stability or immobility increase the area of the base and lower the C.G. as much as is possible. e.g. wrestler wrestling - spread limbs to have greater stability.
3. To stop quickly at rapid motion, spread the base and drop the C.G. as low as constent with the subsequent - movement e.g. in basket ball, dribbling stopping.
4. In all arm support activity the C.G. of the body be as nearly as possible over the point of support on the line of support or on the base of support e.g., hand stand, in gymnastics.
5. Movements of the body when suspended in the air or when hanging by arms or legs follow a definite pattern of movement which tends to maintain equlibrium or body balance.

   e.g., long jump - hang style - **Hip forward**

   spiking in volley ball - **Arching**

   There are two conditions

(A) When the body is free in the air, if the head and feets move up or farward the hips moves to down or backward direction respectively.

(B) When the either hand or feet are supported, if one moves up the hip moves down and vice - versa.

Note – When the body is in the air, the C.G. follows a path where any movement of limbs does not effect it or change the path.

9.2: FRICTION

May be defined as a force that resist or oppose the movement of one body over the other. It acts parallel to the two surfaces and in a direction opposite to that of movement.

(If the either or both the surface are oval the frictional force acts tangentically at the point of contact).

Importance of Friction

Use (1) No friction no movement.

1. Although friction is hinderence to movement but needed for efficient movement –
   e.g., for running faster you need spike.
2. If friction increases results to uncontrolled *motion.*
3. If friction decreases results to falling or slipping

Requirement of friction for efficient movement in particular games or sport varies-

For example: Equipment - basket ball,

Handle of - racket and bat

Surface - synthetic plastic surface for gymnastics.

Shoes - rubber sole, studs, shoes etc.

Factors Deciding Friction

1. The kinds of material which make up the two bodies e.g., shoe surface and playing surface.

2. The condition of the rubbing surface e.g., rougher the surfaces more the friction and vice-versa.
3. Weight of the bodies or the force with which the *surface area* pressed together. Heavier the body more will be the friction and vice versa.
4. The relative action between the two surfaces.

Note:- (Friction do not depend upon the area of the surface come in contact because concentration of weight of the body at a point).

Types of Friction

- Mechanical Type of Friction
- Fluid Friction (H_2O or Air)

Catagories of Mechanical Friction

1. Starting or Static friction.
2. Kinetic or Sliding friction.
3. Rolling friction e.g. ball rolling over the ground.

Coefficient of Friction (C)

Decides the amount of the friction of two surfaces

$$c = \frac{P}{W} = \tan\theta$$

Where, P = Force requiring for movement

W = Weight or pressing force

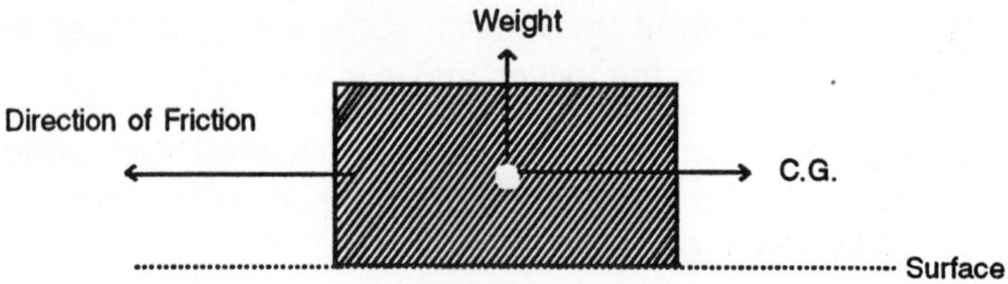

Fig. 9.1:

Direction of Movement

$$\text{Tan } \theta = \frac{\textit{Opposite side}}{\textit{Adjacent side}}$$

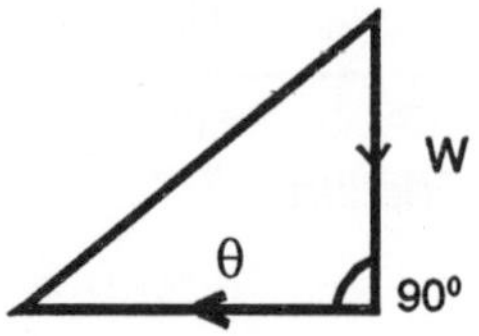

$$= \frac{P}{W}$$

Fig. 9.2:

θ is the ∠ of *friction* also called *limiting angle.*

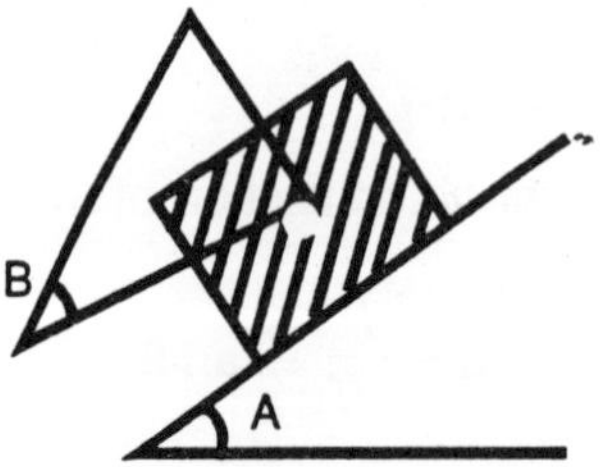

Fig. 9.3:

If A = B then no need to apply *force* to move a body by overcoming it's friction.

Definition of 'C'

The coefficient of friction between two contacting surfaces is the *ratio* of the *force* required to overcome friction to the normal force (usually the weight) pressing the surfaces together, when the surfaces are in rest e.g.

$$C = \frac{\textit{Force necessary to start movement}}{\textit{the force pressing the body together}}.$$

Frictional Forces

As a frictional force always opposes the relative motion of two surfaces, it can sometimes hinder movement and be an unwanted effect. For example as fluid friction, in the form of drag resistance , it hinders the movement of a swimmer through the water and to a lesser degree a diver or high-jumper through the air.

At the other instances, however, it helps movement and becomes a very vital force. For example, consider the foot action of someone walking along a path, which is an example of contact friction. The muscles of the leg push back against the path and this matched by the force of friciton set up between the foot and the path. As a result it is not the foot that moves but the walker's body. If the force of friction is reduced, as it is if the path is icy, the foot may not be held firm and slips back. Then locomotion becomes very difficult indeed.

Contact Friction

This type of friction can be more fully investigated with the apparatus shown in figure below. The load produced by the slotted discs (*P*) will be opposed by the force of friction (*F*) will until the block starts to move, when *F* will have reached its maximum value (*Fm*). Therefore at the point of slipping (P = Fm) and investigation shows that when this occurs, irrespective of the size of the block.

$$F_u = \mu_2 N$$

$$F_u = \mu_2 W$$

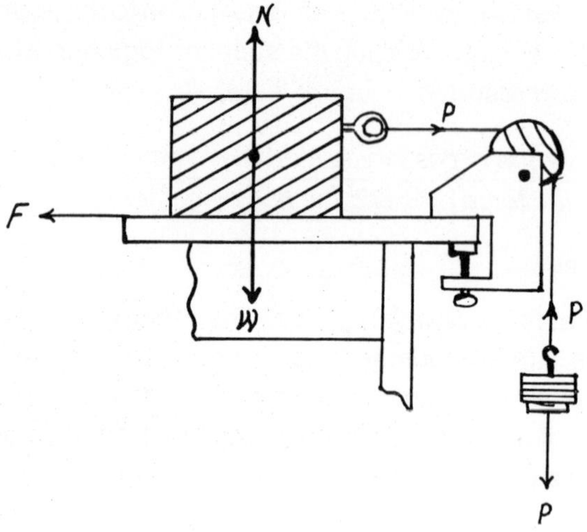

Fig.9.4: The coefficient of static friction

The Fu must be equal to W. The coefficent μ_2, is the coefficient of static friction and is a constant that is determined by the two surfaces in contact. When the block is actually moving, the force of friction drops slightly, but it is still governed by a similar relationship namely:

$$F = \mu_d N$$

Where "μ_d" is of course, less than "μ_2" gives the values of μ_2 for some surfaces which come into constant in human movement studies as followings;

Surface in contact	μ_d
Nylon	015-0-25
Rubber on solids	0-75.4

Wood on wood	0.25-0.5
Leather on wood	0.6
Leather on cork	0.5

It follows that the larger the value μ_2 has for two surfaces, the better the grip. This is very important to manufacture of such things as basket-ball shoes where the better the grip the greater the lean a player can have before be slips over.

Consider the player of given figure. As he pushes into the ground with a thrust (*T*) it will he along force (*F*) will oppose this thrust and therefore be equal and opposite to it. For this condition to be satisfied: resolving horizontally gives us:

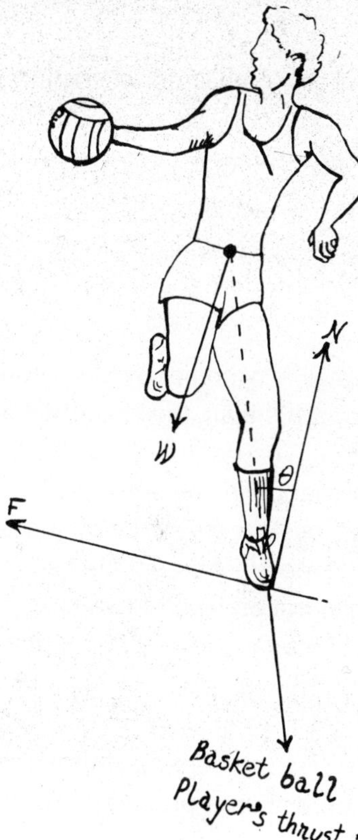

Fig. 9.5: Angle of lean

$$F = T \sin \theta$$

and vertically:

$$N = T \cos \theta$$

Thus

$$\tan \theta = \frac{F}{N}$$

If the player is about to slip, θ will equal θ_m the maximum angle of lean and F will equal F_m which in turn will equal $\mu_2 N$, so that the equation becomes:

$$\tan \theta_m = \mu_2$$

Thus, if the coefficient of friction μ_2 for rubber shoes on a maple floor is 0.75 the angle of lean for a player in this case would be given by:

$$\tan \theta_m = 0.75$$

Drag Friction

Whenever a swimmer moves through the water or an athlete jumps through the air, they experience a resistance force called a drag force. The nature of this force is not easy to explain but it does depend on the effective cross-sectional area that is presented by the swimmer or jumper at right-angles to the direction of their movement their speed, the "streamline" of their body configuration, and the travelling. Dimensionally of the drag force on these factors must have the following form:

$$F_D = KpAv^2$$

Where P is the density of the air or water, A is the effective cross-sectional area presented by the jumper or swimmer and r the velocity of either of them. *K* is a constant dependent on the "Streamline" of their body posture and obviously as neither *K*, *A* nor r remain constant for either the swimmer or the jumper, the drag force they experience is a continually changing quantity. (The details of drag friction will be discussed in fluid friction).

Sports Application of Friction

There are many situations in which athletes try to increase the friction (or "grip") between two surfaces in order to prevent sliding. Softball pitchers cricket baller use resign to improve their grip on the ball; gymnasts use magnesium chalk to improve their grip on the apparatus; and pole-vaulters use sticky adhesive tape and, probably too, a spray grip, resin, or Venice turpentine to achieve a similar result. Basketball boots have soles that have been specially designed to increase the "grip" between shoe and floor, and the pimpled rubber surfaces of table tennis bats ensure a better "grip" between bat and ball than smooth rubber surfaces would allow.

Performers attempt to increase or decrease the "grip" between two bodies by altering the nature of the two surfaces that bear upon each other. In some cases they modify one or both of the surfaces involved (for example, the soles of basketball boots); in others they interpose a substance that has a high limiting friction when in contact with each of the two surfaces and thus has the effect of binding them together (for example, resin and magnesium powder).

From these examples it is apparent that there are two ways in which the friction between bodies can be modified: (1) by altering the nature of the bearing surfaces and (2) by changing the forces that hold these surfaces together.

For two dry surfaces, the limiting friction is equal to the normal reaction multiplied by a constant, the value of this constant depending only on the nature of the surfaces.

that is,

$$F = \mu R$$

where F = the limiting friction, R = the normal reaction, and μ = the constant known as the *coefficient of limiting friction.* Thus, in the case of the lineman and the sled, if the weight of the sled (and therefore the normal reaction) is, say, 1000 N, the coefficient of limiting friction is 0.7.

$$\begin{aligned} F &= \mu R \\ \mu &= \frac{F}{R} \\ &= \frac{700\text{N}}{2000\text{N}} \\ &= 0.7 \end{aligned}$$

The second statement is practically identical to the first and concerns the magnitude of the friction when a body is actually sliding. This is given by the equation

$$F_s = \mu_s R$$

where F_s = the sliding friction, R = the normal reaction, as before, and μ = the *coefficient of sliding friction.* In any given case, the value of this latter coefficient is less than the value for the coefficient of limiting friction. This is in accord with the everyday experience that it is easier to keep a body sliding than it is to start it sliding in the first place.

Rolling Friction

Field hockey and soccer players also study the condition of the playing surface carefully, for they realize that how well the ball rolls across this surface depends very much on how hard, smooth, and dry it is.

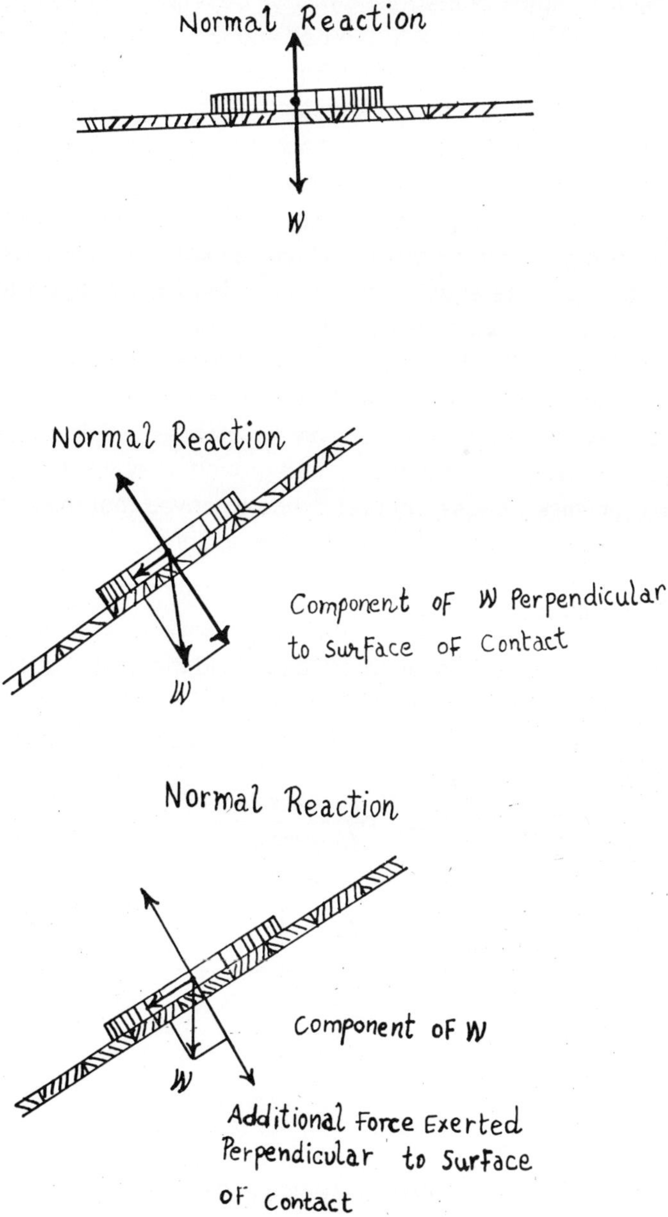

Fig. 9.6: The normal reaction

The magnitude of the rolling friction depends, among other factors, on (1) the nature of the ball and the surface involved, (2) the normal reaction, and (3) the diameter of the ball. The nature of the ball and the surface involved to achieve the best of which they are capable, is to check the prevailing conditions and adjust their games accordingly. If the playing surface is heavily grassed and is soft and wet, the rolling friction will be relatively high and the players will have to exert more force than usual to offset the effect of this high rolling friction. If, on the other hand, the surface is hard and fast, rolling friction will be relatively low and the forces required less than usual. An alternative to modifying the forces exerted is to change completely the tecnique that is used; for example, a field hockey player, involved in a match on a soft ground, may want to use aerial (or "flick") passes in preference to drives that might be stopped prematurely in the "holding" conditions.

Fluid Friction

When an object passes through the fluid media it is offered resistance by their media which called fluid friction. e.g. the fluid friction in case of muscle is *turmed as Viscosity* or *Viscosity of the muscle.*

(a) *Air Resistance* / *(Airodinamics)*

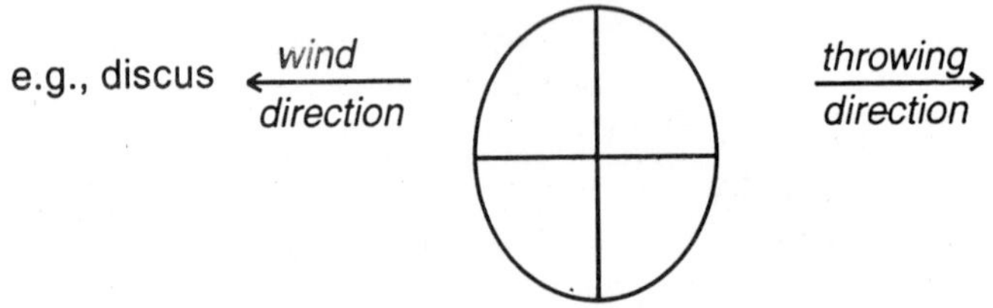

Fig. 9.7:

If held vertically - More resistance in case of head wind due to greater surface area.

e.g., 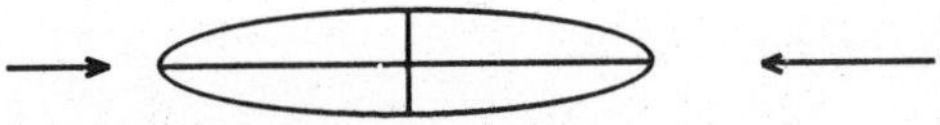 Less resistance

due to lesser surface area.

But practically we do

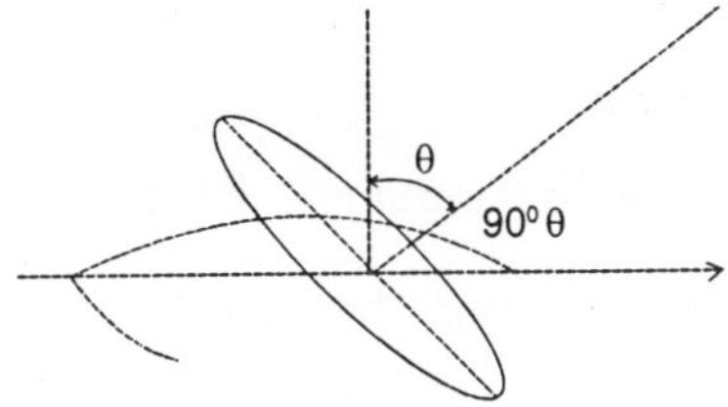

Fig. 9.8:

Has two component reaction is always acting at 90°, making an angle called θ reaction (air resistance) due to the air it acts at 90° or angle

Index of Lifting Efficiency: $= \frac{L}{D}$

Where, L = Lift

D = Drag,

If tell wind increase the surface area to have more drag.

The reaction resulting from the diversion of the air stream about a moving air born body may be expressed by the following equations,

$$R = \frac{ep\, sv^2}{g}$$

Where, R = Resultant reaction (Resistance due to air)

e = A numerical. non dimentional, coefficient depending upon the shape of the body and its attitude to the air stream.

P = Pressure of the air.

S = Active surface area of the body over which air stream passes.

V = Velocity of the air stream with respect to the body (Relative Velocity).

e.g. Cycling

e.g. Wind. Tunnel test.

Water Resistance:- e.g. aquatics, Boating etc.

In Swimming the swimmer has solid body while performs movement through water couses displacement of water thus results into resistance - so body has to do some adjustment such as stream line body.

So to devlop Velocity,

1. Swimmer has to apply optimum muscular force and
2. Reduce the water friction.

Buoyancy

It is the upward force of a liquid on a wholly or a partially immerged body (baised on the principle). "Archemidic stated that a body immerged in a fluid (liquid) is buoyed up by a force, which is equal to the weight of fluid displaced by the body."

As, when a body immerged, if loss of weight is there and this loss of weight is equal to the amount of displaced water's weight.

Specific Gravity

The concept of specific gravity is useful component in swimming.

$$= \frac{\textit{Body weight}}{H_2 0 \textit{ displacement}}$$

= < 1 body will *immerged*

= > 1 body will *float*

To reduce the water resistance the swimmers should keep the following factors in mind, because velocity develops as much by reducing the water resistance as it develops by increasing the force. Therefore, it is desired to reduce the water resistance that the velocity can be built further.

Hence the factors responsible for *building up water resistance* should be controlled, which are as follows;—

(i) Waves
(ii) Eddeis
(iii) Cavitation
(iv) Skin friction
(v) The muscular force
(vi) Starting and stopping
(vii) Internal resistance of body
(viii) Resistance due to the physical features,

According to Carpovitch:– Resistance is proportional to the greater Cross Sectional area of the body.

(Propelling F), $F = Kv^2$

Where, v = Velocity of the swimmer

K = Constant with an average value of 0.6.

Value of 'K' depends on (water resistance due to friction, waves and eddeis).

9.3: IMPACT

Impact is collision or striking of one body with another, causing changes *out come* (movement), a good player should know this concept for *better performance.* e.g. *Football, Tennis, Cricket* etc.

There is a large group of sports in which one body collides (or impacts) with another and in which the success of a participant depends very largely on his (or her) ability to predict the outcome of such impacts.

Tennis and table tennis players to predicting the outcome of an impact between the ball and some part of the playing court, each of these people also faces the problem of what will happen to the ball once it strikes a racquet or bat. In a somewhat different way, golfers, hockey players, and batters in softball or baseball also face this kind of problem, as do athletes kicking in football, heading in soccer, or passing and spiking in volleyball. In view of the prominent role of the impact situation in so many sports, it is important to consider the factors that influence the outcome when two bodies collide.

Coefficient of Restitution

The tendency of a body to return to its normal shape once it has been deformed (that is, its elasticity) differs from one body to another. Some return very quickly to their original shape, while others do so much less quickly.

Newton's Law of Impact

If two bodies move toward each other along the same straight line, the difference between their velocities immediately after impact bears a constant relationship to the difference between their velocities at the moment of impact.

In algebraic terms,

$$v_1 - v_2 = -e(u_1 - u_2)$$

or

$$\frac{v_1 - v_2}{u_1 - u_2} = -e \quad \text{-----------} \quad (a)$$

where v_1 and v_2 = the velocities immediately after impact of bodies 1 and 2, respectively; u_1 and u_1 = their respective velocities immediately before impact: and e = (a) constant known as the *coefficient of restitution.*

Because this law indicates that how bodies move after impact depends on how they were moving before impact and on a coefficient *e,* it is important to consider those factors on which the value of *e* depends. If the ball is called body no. 1 and the floor is body no. 2, the velocities of body no. 2 before and after the impact are both zero for all practical purposes; that is, $u_2 = v_2 = 0$. Equation (a) then reduces to

$$\frac{v_1}{u_1} = -e \quad \text{------} \quad (b)$$

Because velocities are somewhat more difficult to measure than the distances, it is desirable to convert this form of the law into yet another form.

$$u_1 = \sqrt{2gh_d} \quad \text{------} \quad (c)$$

and it can similarly be shown that

$$v_1 = \sqrt{2gh_b} \quad \text{------} \quad (d)$$

where h_d and h_b are, respectively, the height from which the ball was dropped and the height to which it subsequently bounced. If now these expressions are substituted in Eq. (b), this latter becomes

$$e = \sqrt{\frac{h_b}{h_d}} \qquad \text{(e)}$$

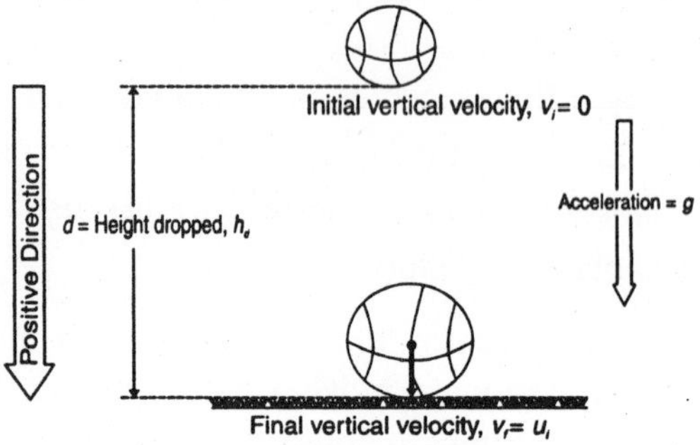

Fig. 9.9:

$$v_f^2 = v_i^2 + 2ad \qquad \text{Eq. (f)}$$

$$\therefore u_i^2 = 0 + 2gh_d$$

$$\boxed{\therefore u_i = \sqrt{2gh_d}}$$

If a ball is dropped from a known height to which it bounces is noted, the value of *e* for that situation can easily be computed. If slight changes are made in the conditions (for example, using a different ball or a different landing surface), the effects that these changes have on the value of *e* can be noted.

Direct and Oblique Impact

There are a few examples in sports where two bodies collide with each other directly along the same straight line immediately prior to impact or one of them is at rest and the other is moving along a line at right angles to the surface so-calld *direct impact* such *direct impacts* are those in which the two bodies do not collide direclty or "head-on." This type of collision is called an *oblique impact.* A bounce pass in basketball is an example of an oblique impact, the ball approaches the floor from an oblique angle. Most of the shots played in racquet games (tennis, squash, etc.) also involve oblique impacts, this means striking it obliquely so that it will not merely return in the same direction from which it came.

Oblique Impact with a Fixed Surface

When a ball strikes the floor it exerts forces on it. These forces can be resolved into components that act along the surface and components that act at right angles to the surface. Fig. 9.10 shows a ball bouncing on the floor in the course of a rally. The velocity it has at the instant it makes contact with the floor of the court is indicated by the vector u and its horizontal and vertical components by u_H and u_H respectively because the ball and the floor are imagined. The horizontal velocity must therefore be the same after impact as it was before, that is, $v_H = u_H$. The same is not true, however of the vertical velocity, for both its direction and magnitude are changed as a result of the impact. In the first place, the force that the force exerts on the ball causes the ball to completely reverse the direction of its vertical motion. Then, in addition, the elasticity of each of the bodies concerned (ball and floor) modifies the magnitude of the vertical velocity in accordance with :

$$\frac{v_v}{u_v} = -e$$

or

$$v_v = -eu_v$$

where v_v equals the vertical velocity of the ball immediately after impact.

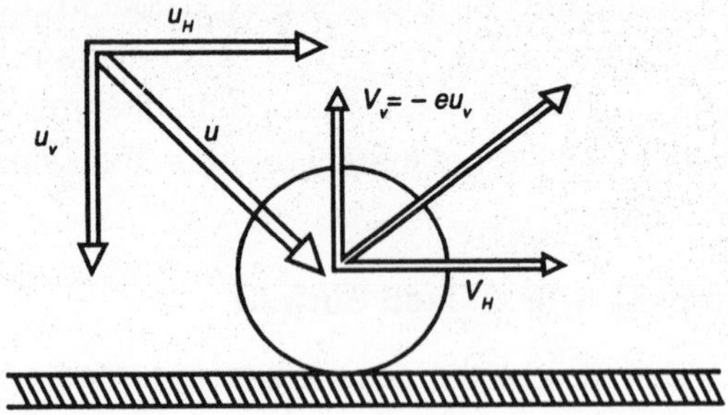

Fig. 9.10: Velocity changes during an oblique impact

By comparing the horizontal and vertical velocities, the resultant velocity after impact can be determined and its direction compared with that of the resultant velocity before impact. For this purpose it is must to define these two directions in terms of the angle that each makes with a line perpendicular to the surface at the point of contact. The angle that the direction of the velocity before impact makes with this perpendicular is called the *angle of incidence,* and the angle that the direction of the velocity after impacts makes with it is called the *angle of reflection.*

Because the horizontal velocities before and after impact are equal in magnitude, and the vertical velocity after impact is less in magnitude than that before impact, the angle of reflection is greater than the angle of incidence. It is important to note that this last statement applies only to oblique impact on a fixed surface, when the effects of friction are negligible or not considered.

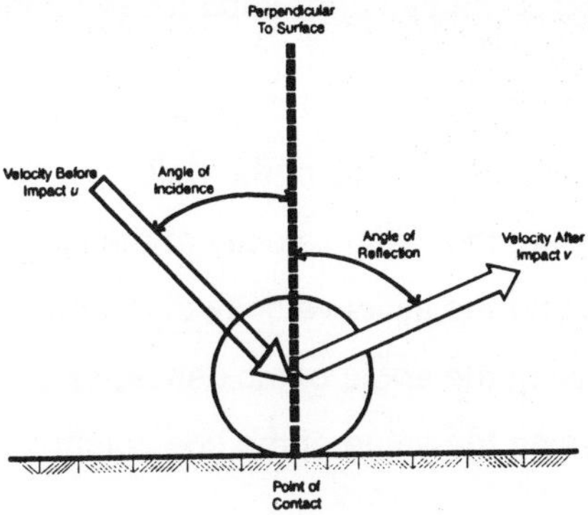

Fig. 9.11: The angles of incidence and reflection in an oblique impact

Oblique Impact with Moving Bodies

Oblique impact between two bodies are those in which both bodies are moving or are free to move. All the cases of hitting, kicking, and heading come within this category that results in two rather involved equations for the speed and direction of a body after such impact. Bat makes an oblique impact with a ball. The velocity vectors u_1 and u_2 represent the velocity of the ball and the bat, respectively. If the acute angle between these vectors is α, the respective masses of the two bodies are m_1 and m_2, and their mutual coefficient of restitution is e, the speed and direction of the ball after impact are given by the following lengthy expressions:

$$v_1 = \sqrt{\left[\frac{m_2 u_2(1+e) + u_1 \cos\alpha(em_2 - m_1)}{m_1 + m_2}\right]^2 + (u_1 \sin\alpha)^2}$$

$$\beta = \arctan\left[\frac{(u_1 \sin\alpha)(m_1 + m_2)}{m_2 u_2(1+e) + u_1 \cos\alpha(em_2 - m_1)}\right]$$

The speed of the ball after impact can be increased by:

1. Increasing the mass of the bat,
2. Decreasing the mass of the ball,
3. Increasing the initial velocity of the bat,
4. Increasing the initial velocity of the ball,
5. Increasing the angle of incidence, or
6. Increasing the value of the coefficient of restitution.

To understand the impact with spin. We take consideration of:

Law of Rebound

This law states if there is no air resistance, no flatting of the ball or equipment and there is no spin imperted to the ball the angle of the rebound after impact with the ground will be equal to the angle of *insident* of the ball.

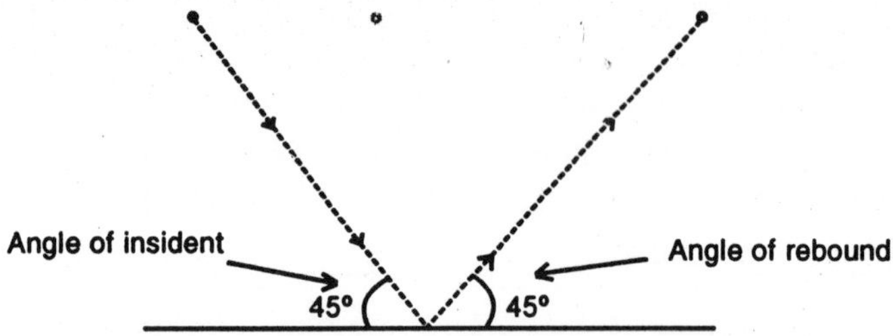

Fig. 9.12:

Restution

The restution of a body is that characteristics, which causes it to resume it's original shape after some force has destarted it.

Elasticity Restitution

It is that quality of a material, when a ball and the surface are slightly compressed. Then since most bodies tend to return to their original shape after they having slightly deformed the ball rebound from the surfaces as the bodies strive to restore themselves to their formal shape. The same sequence of compression and *restitution* takes place, when two moving bodies. e.g. A bat and a ball.

impact

↓

deformation of ball and

↓

the property of ball regain it's *original shape* called *Elasticity.*

Coefficient of Elasticity as Restitution

$$e = \frac{v_2 - v_1}{u_1 - u_2}$$

Where,

v_1 = Velocity of one body after impact

v_2 = Velocity of 2nd body after impact

u_1 = Velocity of one body before impact

u_2 = Velocity of 2nd body before impact

Coefficient of Elasticity

According to Newton's Law of Impact "If two bodies move towards each other along the same straight line the difference between their velocity immediately after impact bears a constant relationship to the differences between the velocity at the moment of impact or compact."

In case of ball and ground (Where ground i.e. one body is in static position) then,

Can also be represented as follows.

$$e = \sqrt{\frac{hb}{hd}}$$

Where,

hb = height of bounce and

hd = height of drop

Law of Conservation of Momentum (For elastic bodies)

"It states that where two or more bodies collide with each other momentum is conserved i.e. the total momentum after impact is equal to the total momentum prior to the impact.

$$m_1 u_1 + m_2 u_2 = m_1 v_1 + m_2 v_2$$

8.4: SPIN

It is the rotation around in any axis and is the result of an application of the *off centre* force.

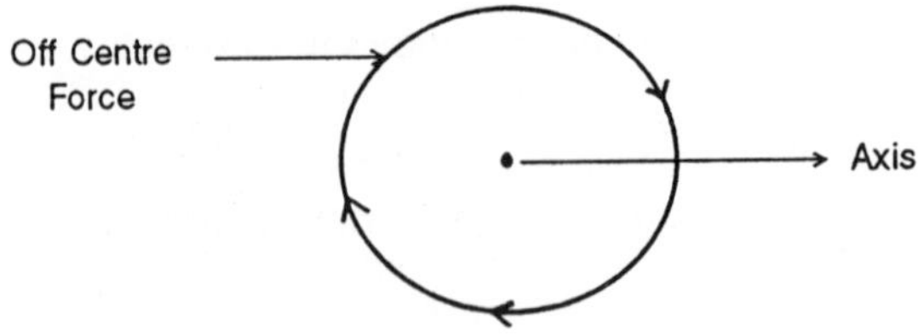

Fig. 9.13 :

Types of Spin

(A) Horizontal Axis Top spin

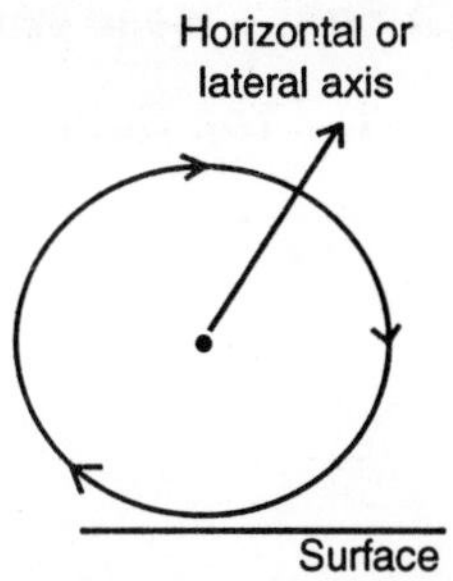

Fig. 9.14 :

Back spin

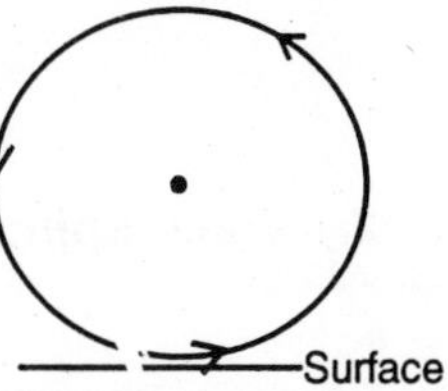

Fig. 9.15:

(B) Vertical Axis

1. Left Spin

(Ball moves towards left side)

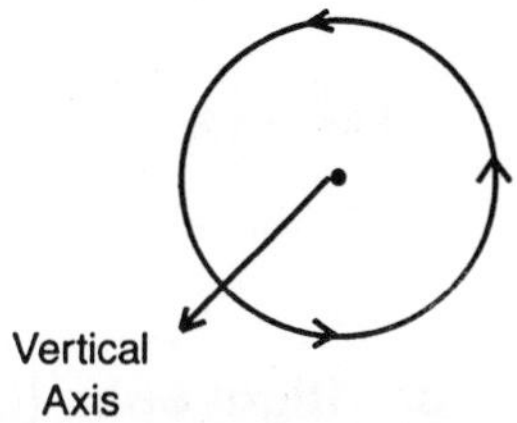

Fig. 9.16:

2. Right Spin

(Ball moves towards right side)

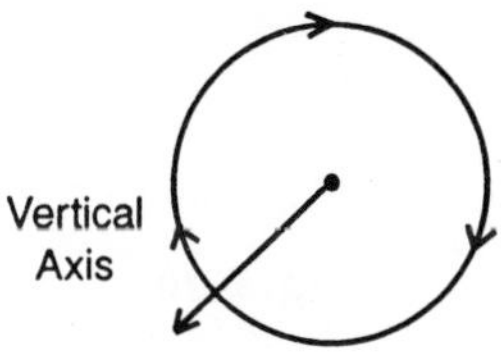

Fig. 9.17:

(C) Spin + Head Wind

1. **Top Spin :** Ball will turndown as resultant movement.

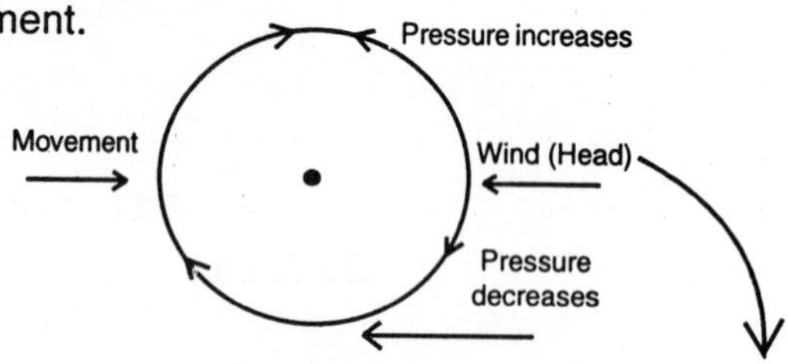

Fig. 9.18:

2. **Back Spin: (ball will turnup Resultant Movement)**

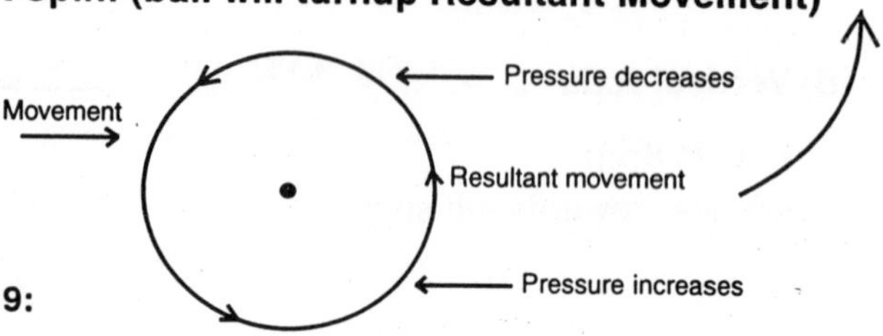

Fig. 9.19:

3. **Right Spin: (Ball will turn to right side)**

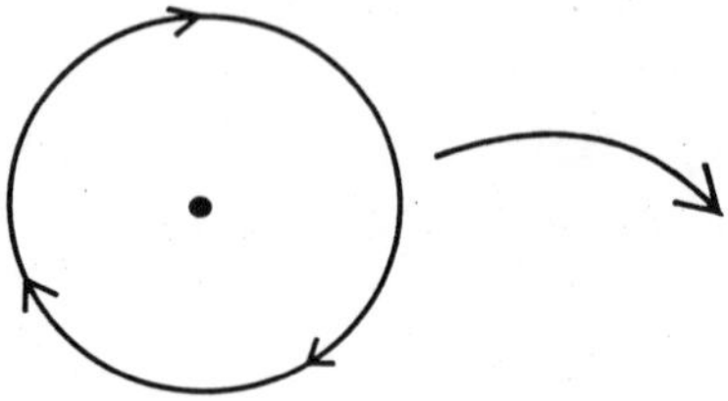

Fig. 9.20:

4. **Left Spin**
   (Ball will turn to left side)

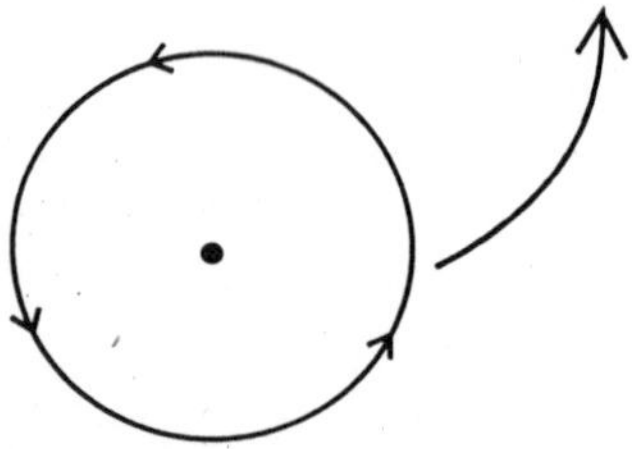

Fig. 9.21:

Effect of Spin on the Angle of Rebound

1. Top Spin + Impact

   90° or

   less than 90°

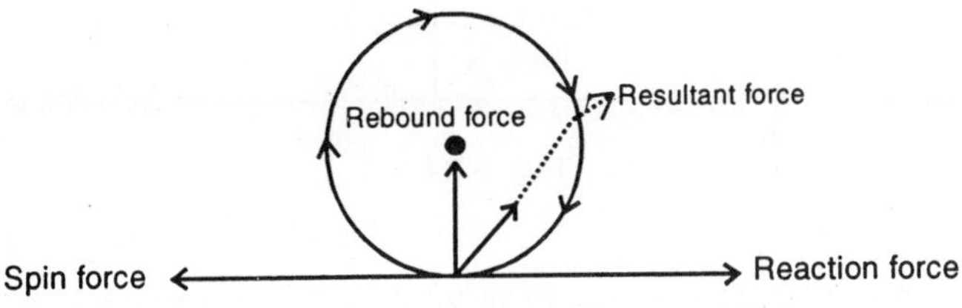

Fig. 9.22:

There are two forces acting together

1. Reaction force
2. Normal Rebound force

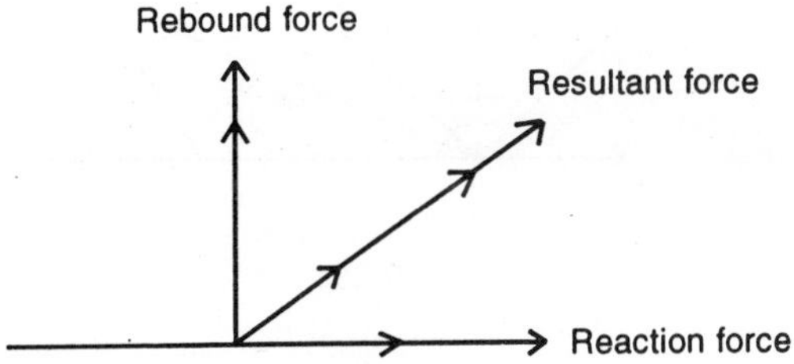

Fig. 9.23

then here, actual path after impact will be resultant of those two forces so it will go away from your body.

2. **Back Spin + Impact** at 90° or less than 90°

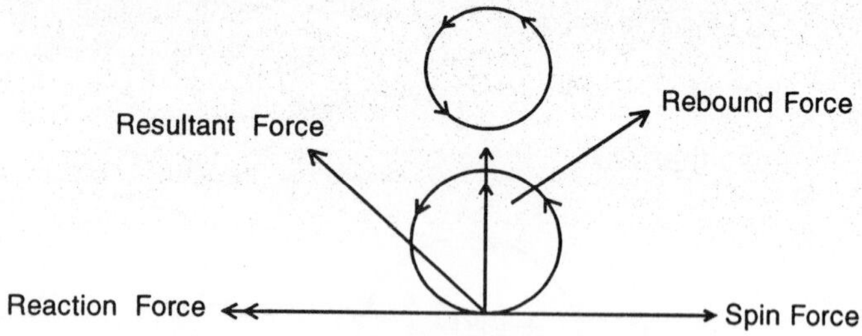

Fig. 9.24

So, it will be coming towards your body after impact to the ground.

Illustrate the Application of Law of Parallelogram of Force When Spin Comes to Impact When Ball Going to Fall in Angle.

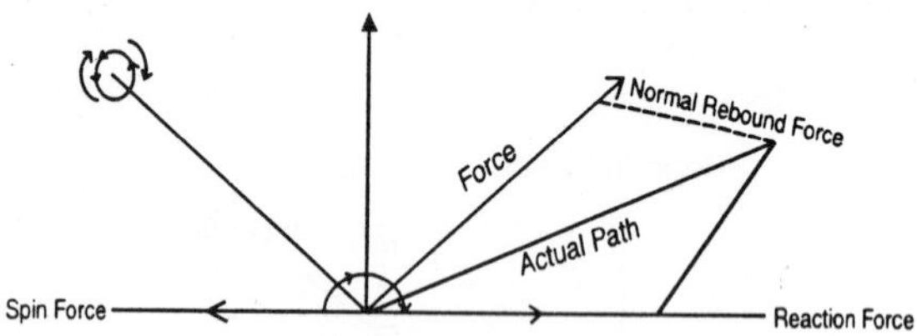

Fig. 9.25:

Reaction force is equal and opposite to spin force, thus after impact the ball on the ground the angle of rebound will decrease i.e. the ball will go *closer* to the ground.

Back Spin

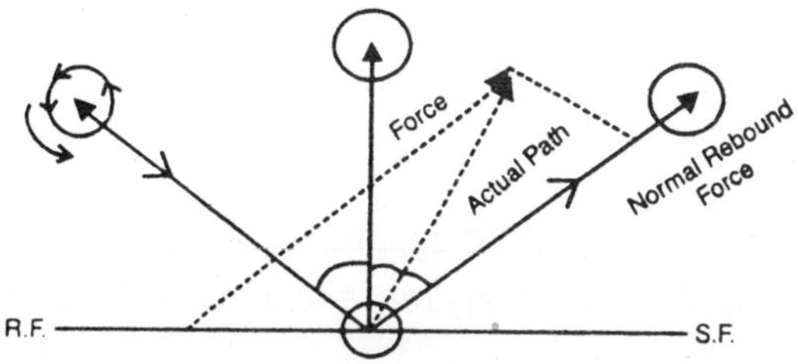

Fig. 9.26:

Thus, due to back spin the angle of rebound will be increased — i.e. ball will go *away* from the ground.

Uses of Spin

1. Spin improves accuracy — Spiner are more accurate than fast or pace bowlers.
2. Spin effects the velocity — with Top Spin the ball — comes to ground quickly, whereas with Back Spin the ball comes slow.
3. Spin provides stability to the implement, when the implement is passing through the space.
4. Flying range or distance are effected by spin.

Swerving

In case of a sportsman running straight while playing but giving a body fake to the opponent is called swerving, which actually involves the horizontal displacement of the C.G. of the body where as the path of the feet is not effected. If a person while running and dodging has to perform a movement around a curve path, the C.G. path is not changed to a curve from the straight line and hence, the person's body moves around two path, one followed by C.G. and other followed by the feet.

9.5: MOMENTUM

M = mass × velocity

Definition

Linear momentum may be defined as the quantity of motion or as the product of moving body mass and the velocity.

Importance of Momentum

1. Momentum is essential for maintenance of velocity e.g. *spirint*
2. Momentum is essential for proper execution of sport skill or technique. e.g. Boxer come close to other boxer. then punches.
3. Momentum is essential to give better performance in case of the jump, and throw.
4. Momentum is necessary for ballestic moment.

Centrifugal Force (CF)

When a body has rotatory motion there develop a force *which tends to cause* the rotating body to *fly off at a tangent* this force is called *Centrifugal Force.*

Conservation of Momentum

According to the impulse-momentum relationship, the respective changes in momentum of the two bodies also be equal and opposite. Newton's first law known as the *principle of conservation of momentum:*

If any system of bodies that exert forces on each other, the total momentum in any direction remains constant unless some external force acts on the system in that direction.

The total momentum is only approximately constant, because the external forces or friction and air resistance are acting.

9.6: FORCE

It can be define as any influence that can change the state of motion of a body. To start or to stop motion, To accelarate or deccelarate, to change direction, essentially this force is either a push or a pull, which may or may not result in motion.

or

Force is the effect that one body exerts on the another body or a body state of being at rest/motion be changed by the action of some other body, the pushing or pulling effects or the other body and that causes the change is termed as force.

"There is No motion without force, but there can be force without motion" According to *Newton's 2nd Law*

$$m \propto \frac{F}{a}$$

$$\text{or } F \propto m.a$$

$$\text{or } F = k\,ma \text{ (when } k \text{ is constant)}$$

F = m.a. If mass remain unchanged where as velocity can be increased

$$F = m\left(\frac{v-u}{+}\right)$$

$$= \frac{mv-mu}{+}$$

$$F = \frac{mv}{f} \quad \text{(When v = average velocity)}$$

Why in Human Body Force is Limited?

1. F = *m*a (mass not possible to change)
2. Acceleration — can be achieved. (By best use of muscular force but limited)

How to Make the Effective Use of Available Force

1. The direction of the force applied.
2. The *point* of force application.
3. The proper sequence in application of generated force (e.g., shot put).
4. The *resultant* effect of the force applied. (e.g., Tug of war, yatching).
5. The time of application of force. (e.g., spiking in volleyball)

   (a) What is time the force is to be applied

   (b) During through which the force applied? (e.g., Wt. lifting)
6. Magnitude of force.

Force and Linear Acceleration

Consider a long board laid on rollers. If it receives a force to the right it will move to the right, if it receives one to the left it will move to the left. If it recieves no force at all it will remain stationary. When a runner crosses the board at a steady velocity we find no net movement of the board. It only moves when he acceleration or decelerates on the board. Thus forces such as those the runner gains from the ground are used to produce either acceleration is required and that is only needed to overcome frictional effects. The magnitude of the acceleration caused by a force will of course depend on the mass of the body being moved. For a given force, the larger the mass the smaller the acceleration. This is why sprinters are lightweights.

Impulse

The shot-putter is performing throw from a specially designed platform that measures the forces he exerts against it. These forces are detected by sensitive strain gauges built into the platform and are then recorded on a long roll of graph paper that is moving at a fixed speed past the recording pens. The horizontal forces that are exerted in the line of the throw — positive forces are those exerted on the ground in the direction of the throw and negative forces are those exerted on the ground in the opposite direction. From this record it can be clearly seen that the magnitude of the force is changing continually during the throw and that the direction of the force also changes.

$$\text{Impulse} = F \times t$$

This product of a constant force and the time during which it acts is also equal to the narrow rectangular area under the force-time curve for that time interval. The total impulse is the sum of the all the infinite number of such similar impulses and can be shown mathematically to be equal to the total area under the force-time curve. The area bounded by the force-time curve and the lines *CB* and *BA* represents the total impulse (300 N.s) in the negative direction, and that bounded by the curve and the line *CD* represents the total impulse (200 N. s) in the positive direction. The algebraic sum of these two values is the total impulse:

$$\text{Total impulse} = (200 - 300) \text{ N.s}$$

$$= -100 \text{ N.s}$$

A useful relationship involving impulse can be obtained by rearranging the algebraic statement of Newton's second law:

$$F = ma$$

Since a the acceleration, has previously been equated with $(v_f - v_t)/t$, this expression can be subsituted for *a* in the previous equation:

$$F = \frac{m(v_f - v_t)}{t}$$

$$F = \frac{mv_f - mv_t}{t}$$

Rearranging this expression yields

$$Ft = mv_f - mv_i$$

The impulse of a force (Ft) is equal to the change of momentum ($mv_f - mv_i$) that it produces.

Pressure

This concept of pressure is of particular importance with respect to safety measures in sports. Performers frequently find it necessary to spread the forces that their bodies must withstand over a large area by initiating a rolling action as soon as they contact the ground. High jumpers and pole-vaulters similarly spread the forces involved when they land flat on their backs on the soft landing pads now available. Landing on the back was thus rarely seen. Instead, athletes in these events generally tried to land on their feet and to cushion the shock as best they could by dropping onto other parts of the body as soon as they landed. Exponents of judo are another group for whom the concept of spreading force over a large area is of fundamental importance. This is readily apparent in the considerable emphasis placed on use of the correct techniques of falling in the early stages of learning the sport. The same basic concept is special protective clothing. Such clothing serves to spread the force over a large area in cases where the same thing is the basis for the use of protective helmets in football and motorcycling; for the wearing of gloves in baseball, cricket, and boxing; and for other similar items of equipment in football (various joint pads), baseball (face masks, chest protectors, etc.), and other sports.

Impulse and Momentum

A force rarely acts instantaneously and we therefore find that for the football kicked in Fig. 9.27 the force experience by the ball changes with time as the force-time graph of Fig. 9.27 (a) shows.

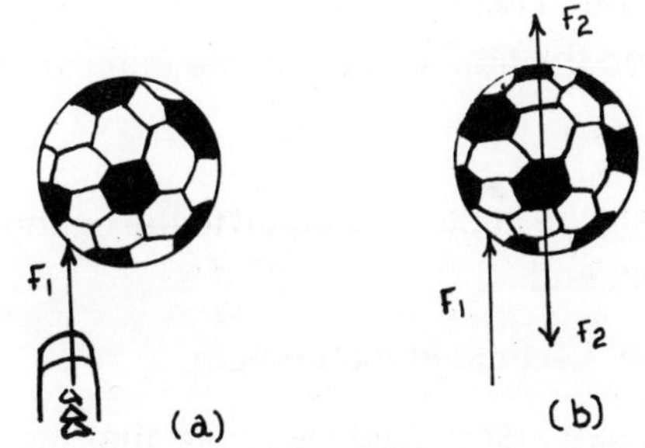

Fig. 9.27-A: A Football Kick

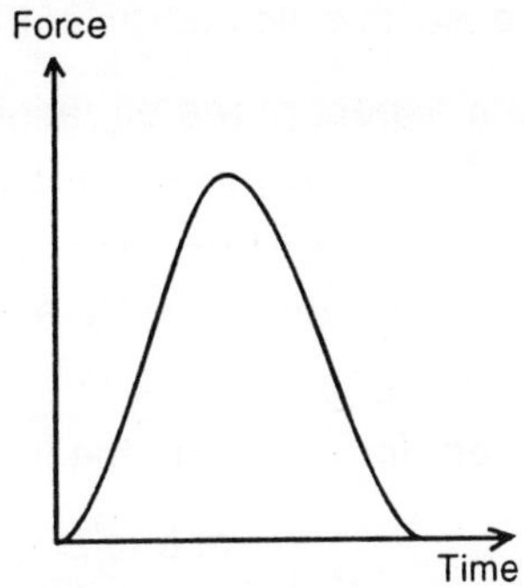

(a) Football kick

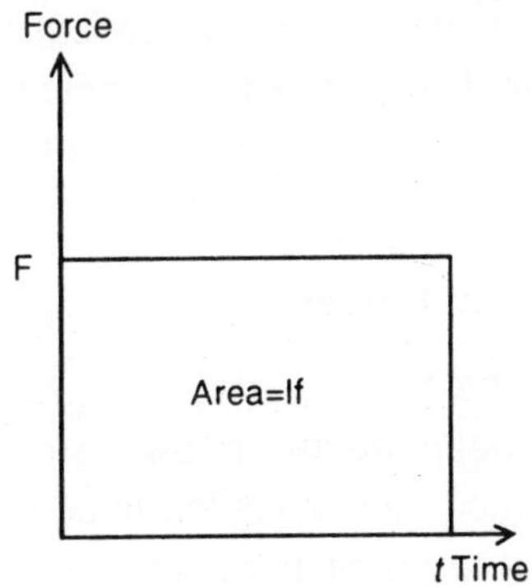

(b) Constant force

Fig. 9.27-B: Force-time graphs

Suppose for convenience a force F acts for a time t on a football of mass m, producing an accleration "a" which changes its velocity from u to r (see Fig. 9.26 (b)). We define the product of a force and the time it acts, as the impulse of a force.

Impulse = Ft

Show that the impulse is equal to the change of momentum that it causes and

Impulse = Change in momentum.

For the case just considered note that the area under the graph is equal to Ft and thus is equal to the impulse and the change in momentum this impulse causes. This is also true for any force time curve, so it would be true of Fig. 9.27 (a) as well.

Another important aspect of the equation comes out when we consider two bodies colliding. When two balls collide the impulse received by ball A from ball B (I_2) is equal and opposite to that received by ball B from ball A (I_1) so that which means that the

Change in momentum = Change in momentum
of ball A of ball B

linear momentum of the system is conserved. This is an important principle and has many applications. It is interesting to consider the implications of this principle of conservation of linear momentum when bat or club hits a ball, and also when a ball is thrown or caught.

Rotational Effect of Forces

See-saw has the turning effect of a force, and the lighter their partner has been the further away from the centre he has had so sit to balance them. This leads us to define the turning effect of a force in terms of its moment which is the product of the force and the perpendicular distance from its line of action to the pivot.

If there is no fixed pivot for a body its centre of gravity acts as the pivot or axis. Thus, if a football is given given a side blow F_1 by introducing two imaginary forces F_2 and F_3 equal in magnitude to F_1 at the centre of gravity of the football we can see that F_1 and F_2 produce a couple of magnitude:

$$G_1 = F_1 X_1$$

which rotates the ball, while F_3 move the ball translationally. the introduction of these two imaginary forces does not change the situation as the resultant of them is zero.

These ideas can probably be better understood by considering their application to a specific human movement situation such as the sprinter.

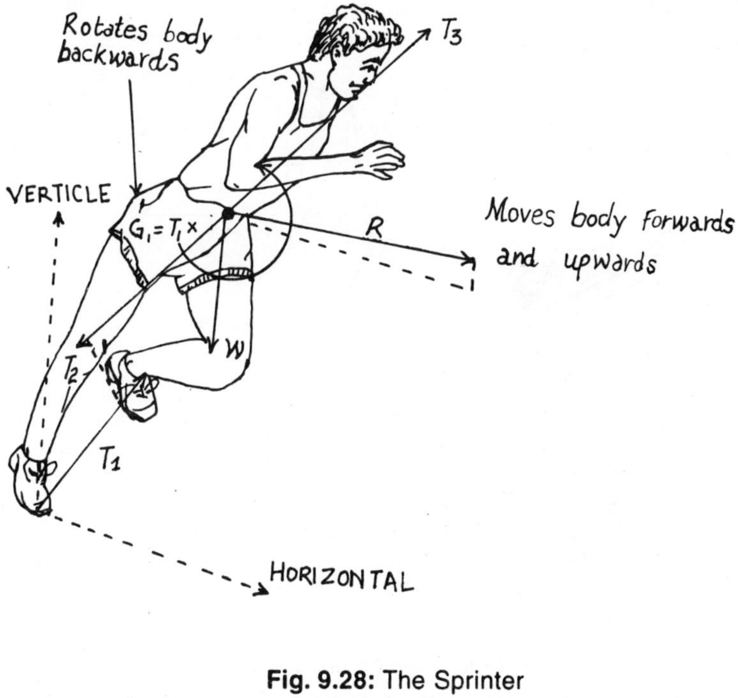

Fig. 9.28: The Sprinter

Fig. 9.28 shows the two forces acting on a sprinter, namely his weight W and the thrust he evokes from the ground T_1. By introducing two opposite forces T_2 and T_3 at his centre of gravity equal in magnitude to T_1 we can see that the couple G_1(of magnitude T_1X) rotates his body backwards, while the force R (the resultant of W and T_3) moves his body forwards and upwards. Only if T_1passes through the sprinter's centre of gravity will his body not be rotated, for X_3 will then be zero. On landing the sprinter's thrust moves from in front of his centre of gravity to behind it, and then moves in front of it again just before he takes off. As a consequence the sprinter's body tends to be rotated backwards, then landing and take-off. In practice the sprinter avoids this "rolling" effect by moving his arms and free leg in such a way as to counterbalance it. How such arm and leg actions do this will be discused in more detail later. Fig. 9.28 shows the two forces acting on a diver, namely his weight Mg and the thrust he experience from the board F_1. In the equivalent

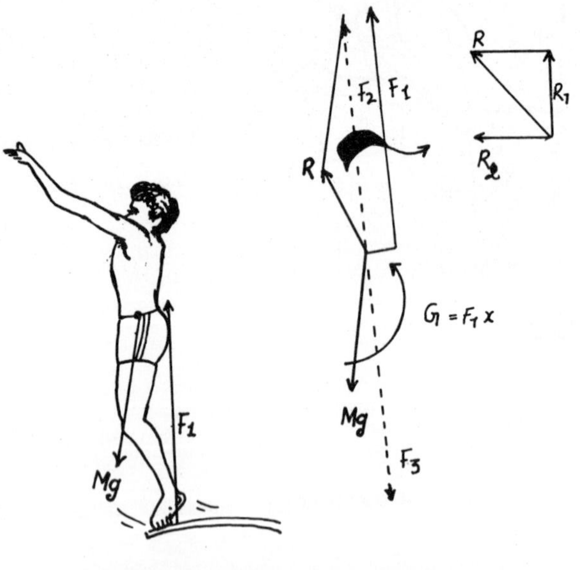

Fig. 9.29: Diver gaining impulse from the board to translate and rotate himself into his dive.

weight *Mg* an the thrust he experience from the boaed F_1. In the equivalent force system F_1, F_2, F_3 it becomes clear that F_1 and F_2 both together to produce a couple *G* of magnitude F_1x that rotates the diver forwards into his dive. Force F_2 on the other hand combines with his weight *Mg* to give a resultant translational force *R*. It is then possible to break R down into a horizontal component R_r which moves the diver horizontally forwards and a vertically upwards. R_r is frequently called the normal contact force and R_x is, of course the frictional force developed between the diver's feet and the diving board.

Force and Circular Motion

If a steel ball is released down it will continue in a straight path. This simple experiment illustrates the first important point about circular motion—to move in a circle an object must be pulled inwards by a force. Such a force is called a centripetal force. Consider the hammer thrower swinging the hammer in Fig. 9.30.

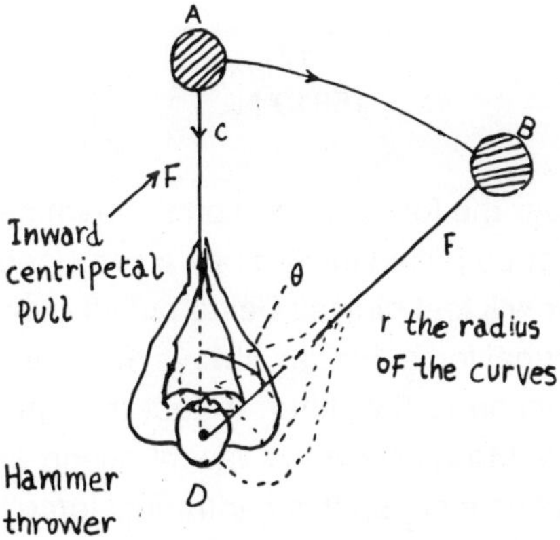

Fig. 9.30: Circular Motion

To an observer the hammer will appear to be accelerating inwards. Let this acceleration be a and the velocity of the hammer be *v*, and let us study the hammer's motion from *A* to *B*. This inward acceleration is of course, the direct product of the inward force and the two are related by Newton's law.

$$F = m\,a$$

As the acceleration towards the centre is equal to r^2/r the inward force involved in circular motion is given by

$$\mathrm{F} = \frac{ms^2}{r}$$

where, *m* is the mass of the object, i.e. the mass of the hammer.

Let us apply this equation to a specific hammer throwing example. The length of the hammer thrower's arm is 0.77 m and the hammer wire I m while the mass of the hammer is 7 k.g. Suppose the hammer thrower lets the hammer go at 20 ms^{-1}. Then at release

$$F = \frac{7 \times 20 \times 20}{1.77}$$
$$= 1581.9 \text{ N}$$

Let us consider the forces acting on the hammer thrower. If the hammer must be pulled inwards by a centripetal force, the hammer thrower will feel himself being pulled outwards by the so-called centrifugal force. Forces always go in pairs like this— one being a reaction to the other. In fact the phenomenon is frequently referred to as Newton's law of action and reaction. To balance the couple set up this centrifugal force the hammer thrower has to lean backwards so his weight can provide a counterbalancing couple. As the speed of the hammer increases, so dies the centripetal force required by the hammer and the centrifugal force experienced by the thrower. To counterbalance this increasing couple he has to lean further back so that the counterbalancing provided by his weight can increase also.

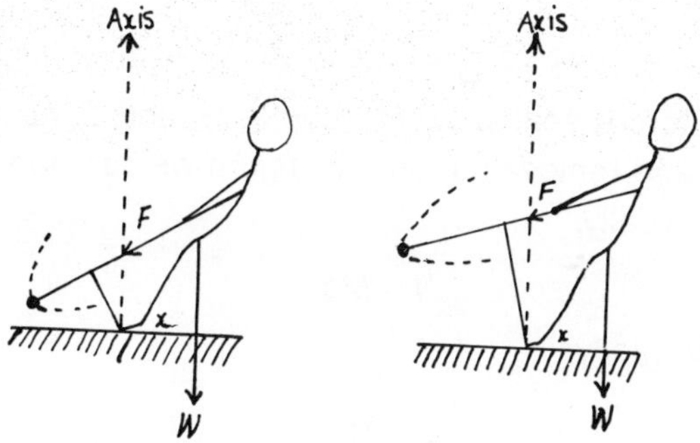

Fig. 9.31: The Hammer Thrower

Human movement contains several examples of circular motion— we shall only consider two of these.

A Runner Cornering on the Flat and on a Bank Running

In order to get the necessary inward force for cornering on the flat a runner has to rely on the force of friction as shown in Fig. 9.32 (a). This frictional force arises as he leans over and, in order not to overbalance the couples set up by the frictional force and the contact force must be equal. Fig. 9.32 (b) shown that for this to be the case

$$\text{so} \quad G_N = G_F$$

$$\text{or} \quad N.\,a = F.\,b$$

$$\frac{F}{N} = \frac{a}{b}$$

but as

$$\tan\theta = \frac{q}{b}$$

this means that

$$\tan\theta = \frac{F}{N}$$

Fig. 9.32 (c) on the other hand shows the translational forces left after the forces setting up the couples G_N and G_F have been considered and removed. From this figure we can see that by resolving vertically.

$$N = Mg$$

and horizontally

$$F = \frac{Mr^2}{r}$$

Thus,

$$\frac{F}{N} = \frac{r^2}{rg}$$

By putting these equations together, we can eliminate F/N and write:

$$v^2 = rg \tan \theta$$

which allows us to predict the lean of a runner for any cornering velocity and cornering radius. If for example velocity is doubled, the radius must be quadupled for the same angle of lean to apply.

However, there is a limit to θ, for the frictional force cannot be larger than to θ, for the frictional force cannot be larger than "μN" where μ is the coefficient of friction between the runner's shoes and the track. Thus, tan θ cannot be larger than m and so a limit is put v^2, namely

$$v^2 \leq \mu rg$$

$$v \leq \sqrt{\mu rg}$$

If a runner tries to exceed this velocity for a given cornering radius, be will have to lean over too far to balance and his feet will slip outward as $G_x > G_F$. On the other hand when cornering on a runner does not need the force of friction as the horizontal component of the contact force can provide enough inward force by itself.

From Fig. 9.32 (d) given above horizontal resolution gives us

$$N \cos (90^0 - \theta) = \frac{Mv^2}{r}$$

$$\text{i.e.,} \quad N \sin \theta = \frac{Mv^2}{r}$$

while vertical resolution gives

$$N \cos \theta = Mg$$

Combining these equations to eleminate N, leaves us with:

$$\tan \theta = \frac{v^2}{rg}$$

or $$v = \sqrt{rg \tan\theta}$$

where v is the velocity that a runner must have to corner on a bank of radius r and slope θ. If his velocity is larger than, his feet tend to slip outwards and the resulting frictional force will, give the necessary extra inward force needed. If his velocity is less than this, the reverse will happen.

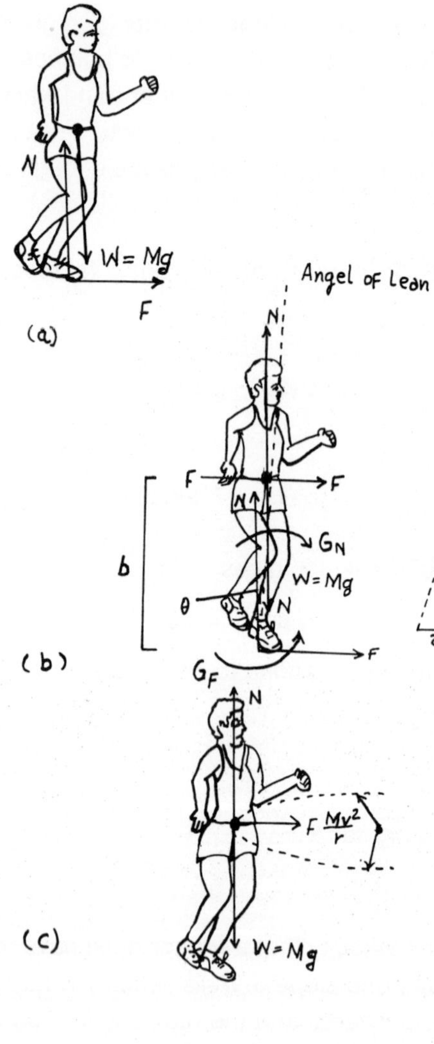

Fig. 9.32: Runner on a Bank

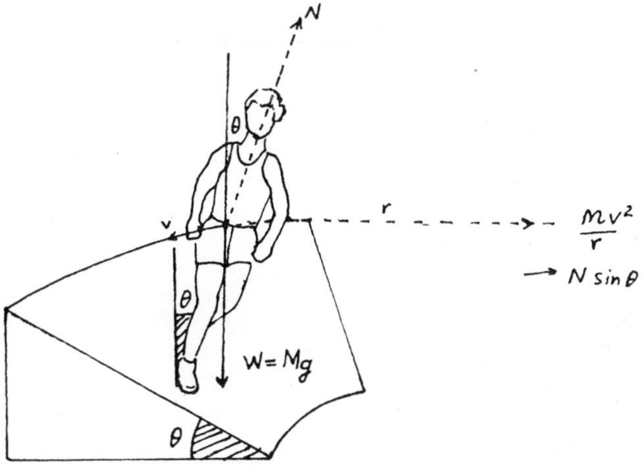

Fig. 9.32-D: Runner cornering on a bank

CENTRIPITAL AND CENTRIFUGAL FORCE

When a body has rotatory motion the body tends to fly off by its tangent called centrigugal force (cf), a force is to be developed tending to pull the rotatory body *towards the Centre of Rotation*. This force is called *Centripital Force (CP).*

Centrifugal Force (CF) is natural in sports technique or events — to nuteralize of *Equal and opposite* cp. has to be developed to counteract the cf. *Equal magnitude and opposite direction If CP and CF balances the movement*

$$cf = m \text{ or } \frac{Wv^2}{gr}$$

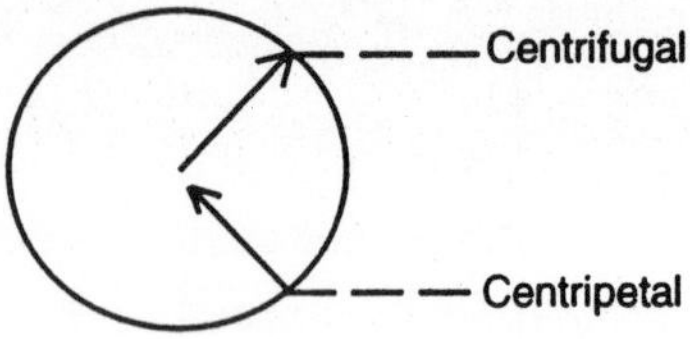

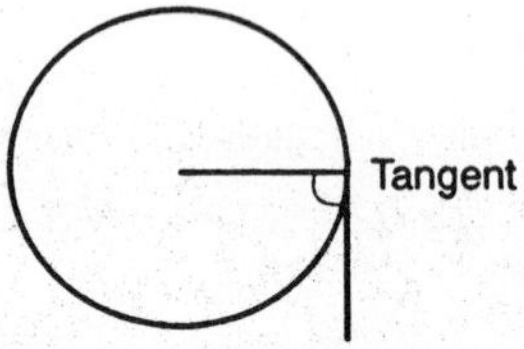

Fig. 9.33:

How to Counteract the CF

1. Leaning towards the centre of rotation. e.g., cycling around a path
2. Raising the bank e.g., Motor race, Hill's road on the couves velodrum.

How to Provide Bank

Tan $\theta = \frac{v^2}{gr}$ To find out the angle through which the runner will lean

Automatic base

3. Pulling towards centre. e.g. Hammer throw, Gaint circle
4. By reducing the radius of rotation e.g., on horizontal bar gaint circle

Contradictial issue = $cf = \frac{mv^2}{g}$ but Vl = Vr Xr

5. By developing optimum friction e.g., in football shoes.

Measurement of Forces

Forces play an important role in human movement studies, let us discuss the methods available for estimating their size and direction of action.

Indirect Measurements

All indirect methods of measuring the force involved in an activity rely on gaining acceleration time graphs from velocity-time graphs and then using Newton's law to calculate the force,

$$F = ma$$

The ticket-tape method for measuring as far as sprinting is concerned is the fact that only the *horizontal* component of the total force is measured. Also unless frames of a cine film can be synchronised to specific points on the tape, it is difficult to relate this force to particular body postures. However, in an activity such as swimming where the net propulsive force is nearly always horizontal and the leg and the arm actions are slower and more well defined, these problems are not so dominant. In this case velocity-time graphs can be produced from water-proofed lengths of tape fed through an a.c. vibrator, one end of which is tied to the swimmer's back. The tape must be kept taut and thus the paying-out spool needs to be effectively braked to prevent the tape sagging. This is even more vital than in the sprint run as the swimmer's motion is more erratic than the spinter's.

The same indirect analysis from velocity-time graphs can be made from stroboscopic photo graphs: for example, the stroboscopic photographs of the tennis serve and golf drive can be analysed to give a velocity-time graph and then an acceleration-time graph from which the net accelerating force acting on the racket or the golf club can be determined. Cine film analysis can also lead to force-time graphs, but care is needed here for very accurate superimposition must be achieved if the results are to be at all reliable.

In this sort of analysis, the motion of the centre of gravity is of prime interest to the researcher because this parameter that acts as the reference points for the translational acceleration and the rotational acceleration that are being produced. Methods of locating the centre of gravity are discussed in the latter chapter which is specifically concerned with the centre of gravity of the human body.

One area where the indirect determination of net forces is particularly susceptible to straightforward computaion is gymnastic activity carried out on the high bar. Fig. shows 8 positions of gymnast doing a giant swing. His centre of gravity in each position is marked by a dot. By estimating the distance travelled from one position to the next by the gymnast's centre of gravity, it is possible to work out his average velocity between two positions. Then the average inward force T his arms are having to exert can be determined by resolving the effect of the forces acting along an inward radius: namely

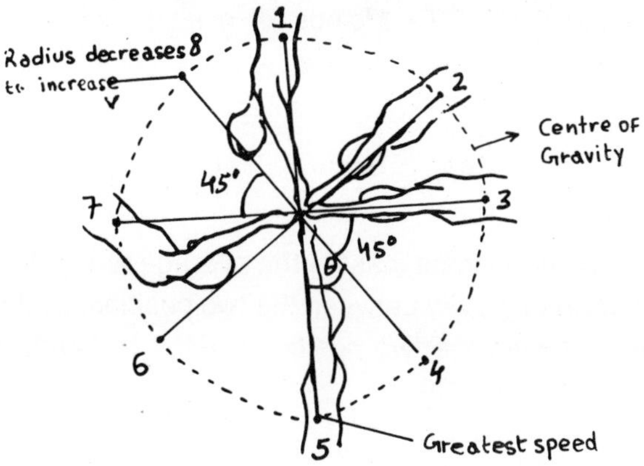

(a) Film analysis speed

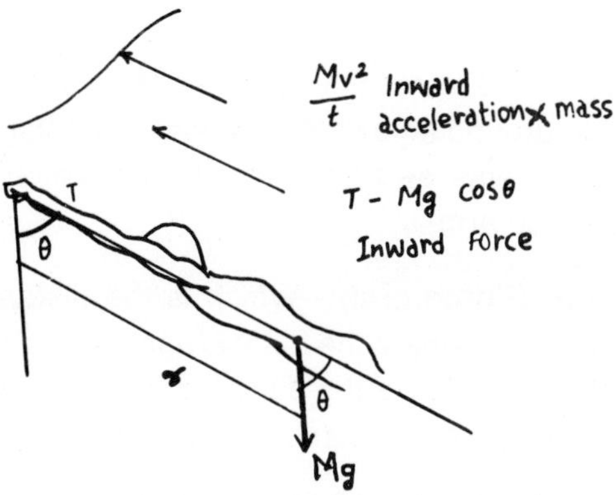

(b) Force -kineton diagram

Fig. 9.34: Giant swing together with a force-kineton diagram relating the inward force T — Mg cosθ to the inward mass × acceleration according to Newton's Second Law F = *ma*.

$$(T - Mg\cos(\theta) = \frac{Mr^2}{r}$$

or

$$T = \frac{Mr^2}{r} + Mg\cos\theta$$

where, is the gymnast mass, r the average radius from the bar to his centre of gravity between the two positions, v the average velocity between the two positions, θ the inclination angle to the downward vertical.

9.7: LEVERS

Definitions

1. Lever is defined as a bar rigid in nature use to over come resistance. Force is applied in other end and weight other end.
2. The lever is a mechanical device to produce turning motion around an axis of rotation. It consist of a fulcrum — which acts as axis of rotation, a power arm and a resistance arm.

A power arm (Force arm):- Which is the distance from the fulcrum to the point where the force is applied and **a resistance arm (weight)** — which is the distance from the fulcrum to the point where weight or resistance is applied.

3. Lever may be defined as a rigid bar (straight or curved) that is used to overcome resistance, when a force is applied to one side of a Fulcrum.

Types of Levers

This is based upon the relative position of the parts of levers that are Fulcrum, Force arm and weight arm.

1. P F W

Δ

Fulcrum is in the middle called *first class lever* etc.

i.e., *seasaw, Hand pump, Bar etc.*

2. ↓ P

Δ W

Second class lever

i.e. Nut cracker

3. P ↓W

Δ ↑

Third class lever

i.e. Forceves, flucker etc.

Principle of Levers (Law or Levers)

When lever is an equilibrium then PXPA = RX RA or FX Force Force Arm (FA) = W × Weight or Resistance Arm (WA)

$$\frac{PA}{RA} = \frac{R}{P}$$

$$i.e., \frac{\text{Lengthof Power Arm}}{\text{Lengthof ResistanceArm}} = \text{Mechanical Advantage (MA)}$$

$$or \; \frac{\text{Resistance}}{\text{Force applied}} = \text{Mechanical Advantage}$$

If the length of power arm is more than the length of resistance arm then the mechanical advantage will be more and vice-versa.

Therefore, mechanical advantage is ratio between the length of power arm and resistance arm.

In Case of Ist Class Levers

The length of *PA* and *RA* is changable — so here mechanical advantage is dynamic. Hence, there may or may not have mechanical advantage.

In Case of II Class Levers

Always has *MA*, which >1

In Case of III Class Levers

Mechanical advantage is < 1 (less than one), so mechanical disadvantage.

What is the Advantage of III Class Lever and How?

Muscles contacts through narrow range but *part* (body part) movement is greater than the narrow movement.

e.g., Hip flexor muscle, while there is contraction there is narrow movement results Flex of thigh (Greater movement) results to speed of movement with greater range.

Range of movement output as efficient locomotion overcoming resistance etc. e.g., of lever of human movement:—

1. Ist class lever e.g., 'V' sit up
2. Floor push up e.g., 2nd class levers.
3. Long seat (extended leg) sit ups e.g., 3rd class levers.

Characteristics of Levers

In Case of Ist Class

1. Can be used for maximum M.A. and vice-versa by shifting fulcrum. e.g., *Triceps* Curl — Ist class lever to build *force* or *strength* or the *speed.*

2. *In Case of 2nd Class Levers*:- M.A. always greater than>1 (greater than one) that is *power arm* is always larger than resistance arm. e.g., Triceps in extension of arms.
3. *In Case of 3rd Class*:- M.A. < 1 because resistance arm is always greater than force arm — Human performance

   e.g., Hip Flexors = Flexion of thigh.

   Hamstring = Flexion of lower leg

   Quadriceps = Extension of lower leg

9.8: MOMENT FORCE OR TORQUE

If there is centric and horizontal force results to no rotation.

Acentric force causes turning motion or rotation.

Torque = T = F × r

Where, F = Force and r = radius or distance between the point (where the force is applied) and axis of rotation.

It is the tendency of a force to rotate a body to which applied about it axis of rotation and it is equal to the magnitude of the force multiplied by the perpendicular distance from the direction of force to its axis or fulcrum.

$$M = FXD$$

e.g., Crouch start, walking

Principle of Moment of Force

In Case of Levers

1. The larger the power arm the greater the moment of force e.g. snatch.
2. The shorter the power arm in proportion into the length of the resistance arm the smaller the moment of force, but more immediate or quicker the action e.g., *Football Kick.*

3. Force has to apply at right angle
   e.g., (i) snatch (ii) Clean and Jerk

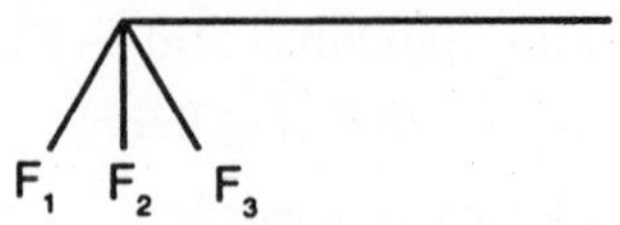

Fig. 9.35:

F_1 and F_3 has two component Horizontal and vertical whereas F_2 has only vertical for best use.

MOMENT OF INERTIA

According to Newton's law of motion the quality of a body to resist change in its state of motion is directly proportional to the mass of that object. The heavier an object the more force, it take to start it moving and more force to stop it in relation to angular motion.

RELATIONSHIP OF TORQUE WITH ANGULAR KINETIC AND KINEMATIC

The torque about any object equals the product of the force magnitude and its perpendicular distance from the direction of force to the point or axis of rotation. This perpendicular distance is called moment arm or torque arm.

The torque or turning effect caused by a five foot pounds weight held in the hand is the product of weight and its perpendicular distance from elbow joint. If this distance is one and half foot the amount of downward turning will be seven and half foot-pounds [Fig. 9.36]. To keep the arm motionless in this position the elbow flexors must exert, equal and opposite or upward directed turning force of seven and half foot-pounds. A force moment is the product of force and moment arm length. It can be increased or decreased by increasing or decreasing either force or moment arm length. [Fig. 9.36].

Rotatory Motion

Fig. 9.36:

The Torque Causing Rotatory Motion

Depending upon the magnitude of force and the perpendicular distance from the line of action of the force to the axis of rotation, the inertia of rotatory body will be affected by distance between the mass and axis of rotation. A ball rotating at the end of a long string is harder to stop and start than one at a short string. As the distance between the axis and mass increases, the inertia increases. This angular inertia is called the moment of inertia (I).

The moment about the axis of rotation is due to inertia and is directly related to the distribution of the mass around the axis of rotation. If the mass of an object is concentrated close to the axis of rotation then object is easier to turn than if the mass is concentrated comparatively away from the axis of rotation. The exact nature of relationship is :

$$I = \sum mr^2$$

Where,

m = Mass of a particle in a rotating body,

r = Perpendicular distance between the mass particle and axis of rotation.

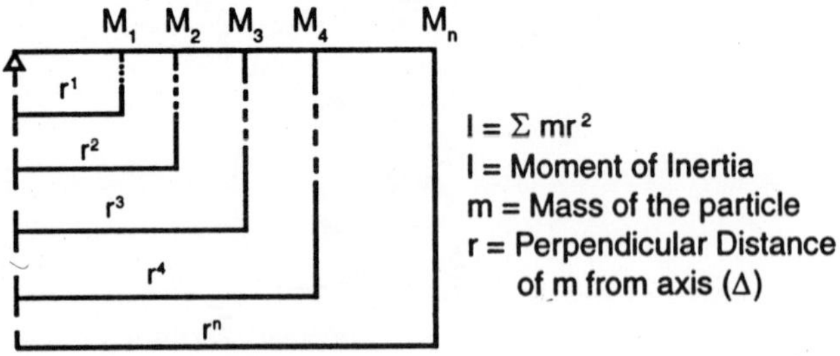

Fig. 9.37:

The moment of inertia about an axis is the sum of mass and square of perpendicular distances between the mass particle and axis of rotation.

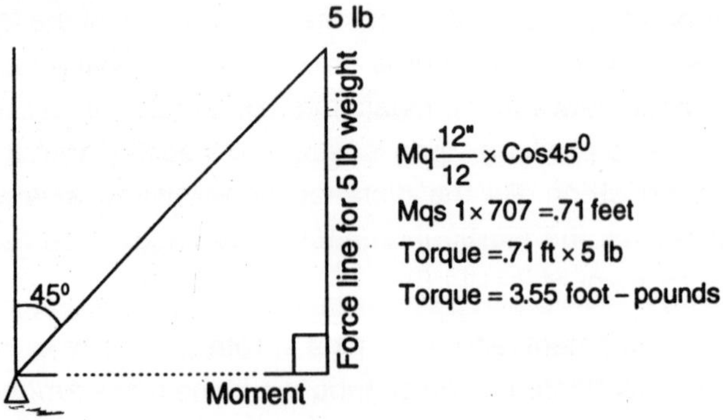

Fig. 9.38:

ROTATORY MOTION

A particle is said to be moving with uniform circular if it moves in a circle with constant speed. When the body is moving in circle there are some forces acting on it. Namely

(i) Centripetal force.

(ii) Centrifugal force.

(iii)Weight of the body acting downwards.

(i) *Centripetal Force:* This force is directed radially inward at every position of the particle. Its magnitude remain constant, though it is changing direction continuously.

Discussion: The centripetal force is an external force acting upon the body from outside and is essentially required to make it move along the circle.

(ii) *Centrifugal Force:* When a body is made to move, uniformly in circular path not only does an external agent exert a centripetal force upon the body, but the rotating body also exerts an equal and opposite force on the agent, tending to move it outwards. This force of reaction tendency to continue moving along a tangential course is called the centrifugal force.

The Human Body in Rotatory Motion

Bending inward of cyclist: A cyclist going round a curve leans inwards to provide the necessary centripetal force so as to be also to go along the curved track. The figure represents a cyclist leaning with the vertical at an angle θ. The force acting on him are :

(i) The weight of the system (mg) acting at the centre of gravity (G) and

(ii) the reaction R due to the ground which in this case, is inclined to the vertical at an angle and also passes through G.

The horizontal component of the reaction, R sin θ provides the centripetal force whereas the vertical component R cos θ, supports the weight. Thus, for a cyclist going round a curve, we have :

$$R\sin\theta = \frac{mv^2}{g}$$

and

$$R\ \mathrm{Cos}\theta = mg$$

It thus follow that greater velocity with which the cyclist goes around the curve or sharper the curve (curve of smaller radius) he takes, the more he must lean inward. For the same reason.

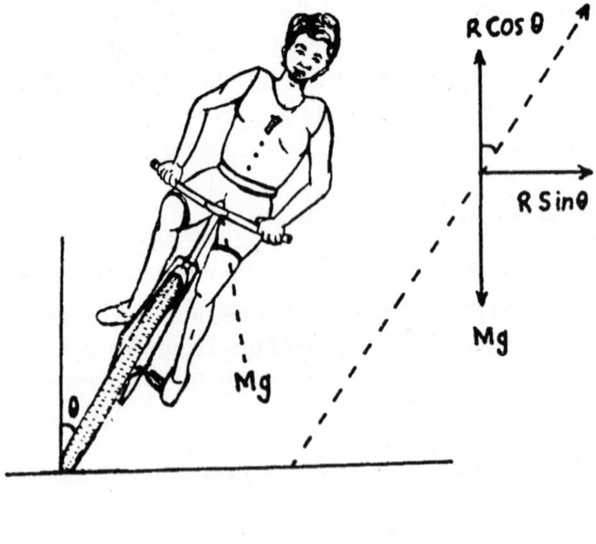

Fig. 9.39:

RELATIONSHIP OF LINEAR AND ANGULAR MOTION

(a) Why throwers take preliminary rotations ?

(b) Why body is crouched in swimming and gymnastic ?

The aim is inhereted in relationship :

$$V_l = V_r \times r \text{ .... } (1)$$
$$D_l = D_r \times r \text{ .... } (2)$$

TORQUE AND ROTATORY MOTION

- Torque and Rotatory motion.
- The effect of two or more torques on a system of vector composition.
- System in linear and rotatory motion.
- Torque and the body's centre of gravity location.

Torque and Rotatory Motion

Whenever an external force acts on a system it causes the body centre of gravity to move in a linear path. (e.g., a thrown ball, pushed opponent etc.). If the system is restricted to moving around an axis, however, the body rotated when the force is applied, only if the force does not act directly through the axis of rotation, that is the point of application of the force must not be at the axis and the line of action of the force must not pass through the axis. Such type of off-axis force is called an eccentric (off-centre) force and the turning effect it has on the body is called torque. The off axis distance to the force's line of action is called the force arm. The greater this radial distance, the greater the torque produced by the force. The specifications of a force arm must be noted however. The force arm is the shortest distance from the axis of rotation to the line of action of force. The shortest distance is always the length of the line that is perpendicular (90°) to the force's line of action (dt.). Perpendicular is designated by the symbol ⊥.

Properties of Torque

All those forces have torque properties they must be considered when dealing with rotating systems. The properties are as followings :

- magnitude (how much force).
- direction (the way the force is applied).
- the line of action.
- the point of application (the point at which the force is being applied to the system).

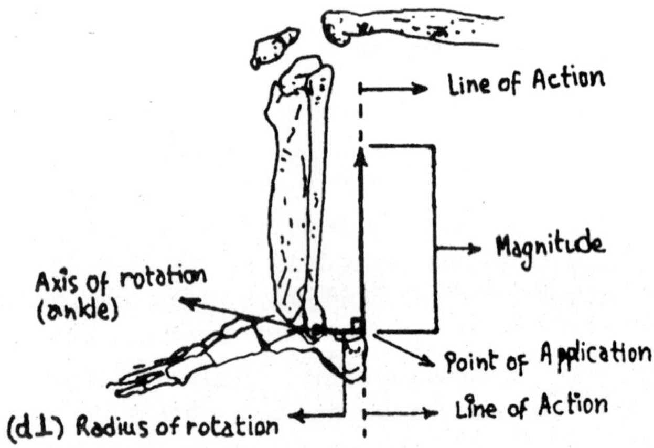

Properties of Torque

Fig. 9.40:

These properties also are associated with torque :

- magnitude of T = F (d⊥)
- directionl clockwise (-) or counter clock wise (+)

The point of application and the line of action of a torque are the same point of application and the line of action as the force producing. The torque must include a force arm (d⊥).

Torque is produced by muscles when they pull the bones, and the result became a rotatory motion of the body segments. The stronger the contraction of a muscle, the greater is the torque on the bones, the longer the force arm, the greater is the torque. Although this point of application of muscle's pull that is in line with the direction of the muscle's shortening changes relative to the bone (the angle of muscle insertion). As the bone rotates, the force arm changes its length. Therefore, the torque that a muscle is able to exert on a bony segment changes as the segment changes its position.

The amount of torque acting to rotate a system is found by mutiplying the magnitude of the applied force by the force arm distance from the line of force to the axis of rotation. Where, T is torque (Newton meters or Pound feet), F is force (Newton's or Pound) and d⊥ is the length (metres or feet) of the shortest line drawn form the axis to where it meets the line of action of the applied force at 90^0.

$$\text{Torque} = \text{Force} \times \text{force arm or } T = F\,(d\perp)$$

If a force of 50 N were applied with a force arm of 2 m. The torque produced by that force would be 100 Nm or 100 torque units. The larger the force and/or the longer the force arm, the larger the torque. For this reason, with two equal forces applied, the force that has the longer force aim produces the larger torque.

The Effect of two or more Torques on a System (Vector Composition)

When more than one external torque is applied to a system at the same time all the torques must be considered for their net or resultant effect determines the rotatory motion of the body.

The two applied torques in Fig. 9.14 tends to rotate the segment in opposite direction they oppose to one another a counter clockwise (ccw) direction is called the positive direction of rotation and a clockwise direction (cw) called the negative direction. The rotatory motion response of the segment is determined by calculating the net torque is similar to finding the net force on a system and they may be found by algebraically adding. The two or more applied torques, with plus and minus signs used to indicate the direction of each.

(+70 torque units) + (–27 torque units)
= +43 torque units as resultant,

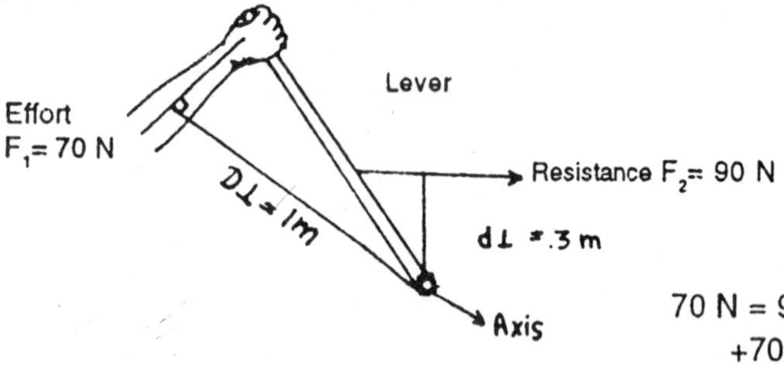

$$70\text{ N} = 90 \times .3$$
$$+70 = -27$$
$$T = (70 - 27) = 43$$

Fig. 9.41:

System in Linear and Rotatory Motion

In purely linear motion of a systems the centre of gravity is caused by a net external force whose line of action passes through the body's CG. Purely rotatory motion of a body occurs when a torque or offcentric force/forces is/are applied on a body with a fixed axis. Such is the case in many sports and particularly activities in which a system is projected into the air. Any unsupported rotatory system rotates around an axis of rotation that passes through its C.G.

Example :

A volley ball served with back spin is shown in fig. 9.14. The balls C.G. travels along a curvilinear path. At the same time, the ball rotates around an axis passing through the C.G. When a force has been applied eccentric to the axis of rotation to produce these linear and rotatory motions.

During the time that the striking hand is in contact with the volley ball. The force of the hand acts forward and upward and eccentric to the axis of rotation at C.G. Thus, this force causes initial upward and forward linear motion of the ball C.G.

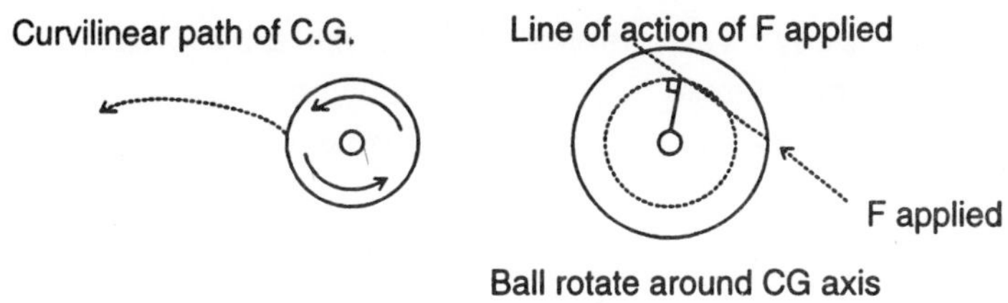

Fig. 9.42:

If the line of action of the hands force passes through the axis of rotation, no torque is produced and therefore no spin.

Torque and the Body's Centre of Gravity Location

As the centre of gravity (C.G.) is frequently the point in a system that is used for motion analysis. The C.G. of any body is that point through which the force of gravity acts and all the body's weight seems to be concentrated at this point. The magnitude of the force of gravity acting on the body is called the weight of that body; its line of action is vertical, towards the centre of the earth. The force of gravity actually pulls on every particle of mass in the body and thus gives each part weight. For an understanding, it is useful to consider the individual moveable body segments as the smallest divisions of the total body's mass. Each body segment or mechanical link, has its own mass or weight, and C.G. and the gravity exerts a downward force through each partial C.G.

The C.G. of each segment is situated closer to each segments weight which influences not only the total body weight but also the total body's location. One definition of the C.G. of a system is that the C.G. is an imaginary point about which the body weight is equally distributed that is point of balance. According to the definition, the sum of the segmental weight torques on all sides of the body's C.G. point must be equal to zero. The body's C.G. serves as the point around which the body segments would balance. The body parts would balance on either side of the C.G. without regard to the body's orientation in space.

The C.G. location is an important factor in movement analysis and therefore we should be able to locate the C.G. of the total body with the body segments in different positions. Since the segments weights do not change during movement. The positions of the segments relative to one another is the important factor in producing a change in the segmental torques and thereby the location of the C.G. of the body.

Vector Resolution of Angular Momentum

Angular Momentum

Angular momentum is the product of the angular equivalents of mass and velocity that is the moment of inertia (I) and angular velocity (w).

∴ Angular Momentum (L) = Iw .... (1)

Vector Resolution of Angular Momentum

We consider a particle of Mass (m) moving with a velocity (v), Let at any instant the position of the origin A be specified by the position vector (r). We are considering the general case in which the position vector (r) is inclined to the three axis and is making different angles with them. Let the linear momentum of the particle (M = mv) be making an angle θ with the position vector (r). The vector M can be resolved into two components: (i) M Sinθ perpendicular to (r) and (ii) M Cosθ parallel to (r). The moment of momentum of the particle about the origin A is called the angular momentum of the particle about the A. It is obvious that the momentum of the angular momentum and its direction is anticlockwise. It can be represented by a vector L in the direction. So in the vector rotation the angular momentum of the particle about the fixed origin is defined by the equation; L= r × m

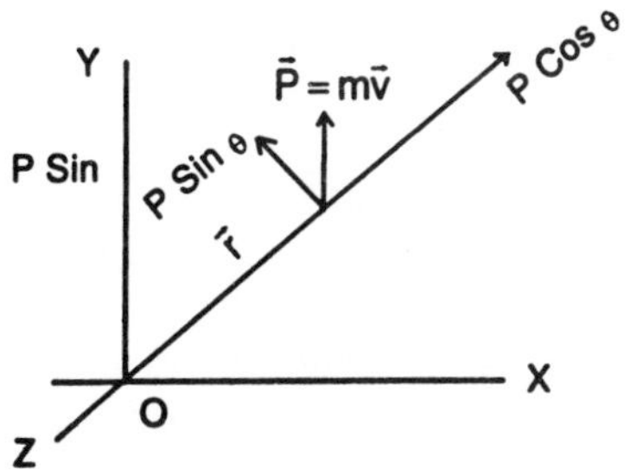

Fig. 9.43:

9.9: WORK

Work is said to be done if a force when applied is able to move the body through certain distance.

$W = F.D$ Where,

or $WD = F.D$

W = Work done.
F = Force applied.
D = Distance covered

running, jogging, walking in the spot - No *W* done,

Useful work done = WD = $F \times h$ (e.g. high jump)
e.g., Stair climbing WD = $F \times d \sin \theta$ (e.g. in case of two type of work done)

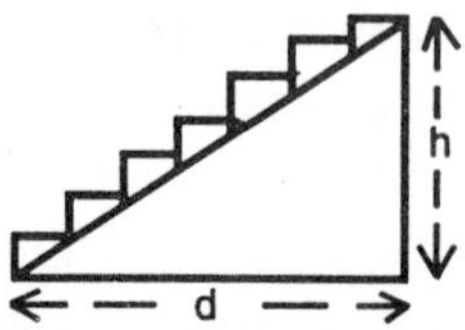

Fig. 9.44:

Consider a weight lifter performing a two-hand snatch, a lift in which a barbell is raised over head in one uninterrupted movement. If he exerts a constant upward force of 1800 N against the barbell, while lifting it the first 0.5 m from the floor, the work done by the upward force during this portion of the lift is

$$W = 1800 \text{ N} \times 0.5 \text{ m}.$$
$$= 900 \text{ N.m (or } \textit{joules)}$$

When a person does a pull-up on a horizontal bar, the muscles of the arms and shoulders provide a vertical force that lifts the body. Because the direction in which the body moves is the same as that in which the force exerted by the muscles acts, these muscles can be said to do positive work. When a person then lowers the body, these same muscles exert an upward force to help control the descent. During the descent these muscles are doing negative work.

9.10: POWER

"The rate of doing work" or "The work done per unit times".

$P = \frac{W}{t}$ faster the work done more is the power required and vice-versa.

Sprint — as Power event

$$P = \frac{W}{t} \quad \text{or} \quad P = \frac{F \times d}{t}$$

$$P = \frac{Fxd}{t}$$

or P = F. V

V = average velocity

e.g., 1 mile run

A-4:26 Pace runner (constant)

B-4:26 Varying pace

To determine who is applying better power A or B;

One principle applied $P = (v)^3$. 3 here better Power is applied by B.

Power's Definition with Examples

If a weight lifter does say, 1500 J of work in raising a barbell overhead, the amount of work is in no way dependent on how long he (or she) took to perform the feat. Whether the lift took 0.5 sec. 1 s, or even 2 s, the amount of work done is still

1500 J. What does change though is the *power,* or the rate at which the work is performed, for to arrive at this value the work done is divided by the time taken.

9.11: ENERGY

"Can be defined as the potentiality to do work or the ability to perform work".

(Performance is better when there is economy of energy in movement for an efficient person).

Energy (E)

Kinetic E (KE)
(Energy due to motion)

Potential E (PE)
(Due to change in position or deformation in shape of the object).

$KE = \frac{1}{2} mv^2$

PE = mgh or (Wh)

e.g., Raising high jump pitch, Tampoline Deformacy,

Catching cricket ball — Ping-Pong-ball-Elasticity, storing water in the water tank in certain height.

Law of Conservation of Energy

Energy neither be created nor be destroyed. Change in state/ form. In given from one movement to another movement. i.e., change from *KE to PE – thus E* remain constant.

Relation PE with — KE

$KE = \frac{1}{2} mv^2 = W = FD$

W = F.D

$$W = \frac{ma \times v^2}{2a} = \frac{1}{2}mv^2 = KE, \therefore v^2 = 2as \text{ or } s = \frac{v^2}{2a}$$

According Newton's Law:—

$F = ma$

$v^2 = 2as$

or $v^2 = 2ad$ or $2gh$

and $= \dfrac{v^2}{2a}$

$KE = \dfrac{1}{2}mv^2$

$= \dfrac{1}{2}m\,zgh$

$= mgh = PE$

e.g., Freely falling body

If $KE = W = F.D$

"The capacity to do work" and this latter quantity is measured in joules, it is convenient to use the same unit in measuring all three kinds of mechanical energy. This change in the unit used is greatly helped by the fact that

$1 \text{ kg. } (m/s)^2 = 1 \text{ N. m}$

$= 1 \text{ J}$

Therefore, whenever the kinetic energy of a body is being computed, and the mass is expressed in kilograms and the velocity in meters per second, the answer can be taken to be a joules.

Potential Energy And Kinetic Energy For Sports Performance

A trampolinist at the peak of his flight has capacity to do work because of his position relative to the surface of the earth. The energy due to the position that a body occupies relative to the earth's surface is called *potential energy* and can be determined by multiplying the weight of the body (the force) by its height above the surface (the distance):

$$E_p = Wh$$

where E_p = the potential energy, W = the weight of the body, and h = its height above the ground.

The changes in kinetic and potential energies experienced by a trampolinist must also be experienced by any body that becomes airborne, and confirmation of this is contained in the *law of conservation of mechanical energy*, which, for present purposes, may be stated as :

When gravity is the only external force acting on it, the mechanical energy of a body is constant.

Thus, when a body is in flight (and the effects of air resistance are small enough to be ignored), the sum of the body's kinetic and potential energies is constant.

Work Energy Relationship

There is an important relationship among work kinetic energy, and potential energy. Consider an arrow about to be fired horizontally from a bow and assume that a constant horizontal force is exerted on the arrow from the moment the archer releases it until it loses contact with the string. Immediately before the bowstring is released, the arrow is at rest and therefore possesses no kinetic energy. Then, when the archer releases the bowstring, work is done on the arrow until it loses contact with the string, by which time it has come to possess a certain amount of kinetic energy. The average acceleration of the arrow is $v_f^2/2d$. Substituting this term into the equation of Newton's Second Law yields,

$$F = \frac{mv_f^2}{2d}$$

where F equals the horizontal force exerted on the arrow. If this equation is rearranged,

$$FD = \frac{1}{2}mv_f^2$$

The left-hand side is equal to the work done on the arrow by the bowstring, and the right-hand side is the kinetic energy that the arrow possesses as it leaves the bow.

When a ball is being caught, kinetic energy is used to do work on the hands of the person who catches it. Example, if a ball does 80 J of work against a person's hands, the force exerted is 800 N if the hands move 0.1 m:

$$W = Fd$$

$$F = \frac{W}{d}$$

$$= \frac{80}{.1}$$

$$= 800 \text{ N}$$

and only 160 N if the hands move 0.5 m:

$$F = \frac{80}{0.5}$$

$$= 160 \text{ N}$$

Consider again the bow and arrow and imagine the arrow directed vertically upward rather than horizontally. Assume too that a constant vertical force is exerted on the arrow from the instant the archer releases it until the instant it leaves the bowstring. If the speed which the arrow leaves the bowstring is v_t· the acceleration of the arrow during the period that the bowstring is applying force to it is $v_f^2 / 2d$, as the before. The force exerted on the arrow during the same period is equal to $F - W$, where F is the upward vertical force exerted via the bowstring and W is weight of the arrow.

$$F - W = \frac{mv_f^2}{2d}$$

and rearranged,

$$(F - W)d = \frac{1}{2}mv_f^2$$

$$Fd = \frac{1}{2}mv_f^2 + Wd$$

If can be seen that the work done is equal to the sum of the changes in kinetic and potential energy that result. (This is the so-called *work-energy relationship).* Thus, for example, if a 800 N basketball player raised his body 40 cm. and acquired a velocity of 4.5 m/s by the time she left ground in a center jump, the work done by the muscles producing this motion could be computed as follows :

$$\begin{aligned} \text{Work} &= \left(\frac{1}{2} \times \frac{800}{9.81} \times 4.5^2\right) + (800 \times 0.4) \\ &= 825.68808 + 320 \\ &= 1145.688 \text{ J} \end{aligned}$$

The work done by a weight lifter in raising a 800 N barbell at a distance of 1.8 m. to a stationary position overhead (that is $v_f = 0$) can be computed :

$$\text{Work done by lifter on barbell} = \left(\frac{1}{2} \times \frac{800}{9.81} \times 0\right) + (800 \times 1.8)$$

$$= 0 + 1440$$

$$= 1440 \text{ J}$$

If the athlete himself weighs 600 N and raises his body 0.7m. in the course of getting the barbell aloft, the total work done by the lifter is

Work done on barbell + work done on lifter = 1440 + 420

= 1860 J

Chapter – 10

TWO DIMENSIONAL ANALYSIS

When a body is moving in a plane, the position of the body can be determined if its co-ordinates (x, y) measured parallel to two fixed perpendicular lines $\overleftrightarrow{OX}, \overleftrightarrow{OY}$ in the plane be known.

Let the body be moving along the curve APQ, and let it move from the position P at time t to the neighbouring position Q at time $t + \Delta t$.

Let the co-ordinates of P be (x, y) and those of Q be $(x + \Delta x, y + \Delta y)$ with reference to the rectangular axes $\overleftrightarrow{OX}\ \overleftrightarrow{OY}$.

The displacement of the body parallel to $\overleftrightarrow{OX}$ in time Δt is Δx.

$\therefore$ the resolved part of the velocity parallel to $\overrightarrow{OX}$, namely,

$$V_x = \underset{\Delta t \to 0}{Lt} \frac{\Delta x}{\Delta t} = \frac{dx}{dt}$$

Similarly, the resolved part of the velocity parallel to $\overrightarrow{OY}$, namely,

$$V_y = \underset{\Delta t \to 0}{Lt} \frac{\Delta y}{\Delta t} = \frac{dy}{dt}.$$

Hence, if (x, y) be the co-oridinates of the point P at any instant with reference to a set of rectangular axes $\overleftrightarrow{OX}, \overleftrightarrow{OY}$, the component velocities parallel to the axis are the rates of change of x and y.

$$\text{Thus, } v_x = \frac{dx}{dt} \qquad \ldots \qquad \ldots (1)$$

$$v_y = \frac{dy}{dt}. \qquad \ldots \qquad \ldots (2)$$

The resultant velocity V is given by

$$V = \sqrt{\left(vx^2 + vy^2\right)} = \sqrt{\left\{\left(\frac{dx}{dt}\right)^2 + \left(\frac{dy}{dt}\right)^2\right\}} \quad \ldots \quad \ldots (3)$$

Also, if θ be the angle which the direction of the resultant velocity (*i.e.*, the direction of motion) makes with the x-axis, we have

$$\tan\theta = \frac{dy/dt}{dx/dt} = \frac{dy}{dx},$$

i.e., the direction of motion at any point is the same as the direction of the tangent to the path at the point.

Similarly, component of acceleration f_x and f_y along $\vec{OX}$, $\vec{OY}$ being the rates of changes of velocities along these directions. we have

$$f_x = \frac{d}{dt}(v_x) = \frac{d}{dt}\left(\frac{dx}{dt}\right) = \frac{d^2x}{dt^2} \quad \ldots \quad \ldots (4)$$

$$f_y = \frac{d}{dt}(v_y) = \frac{d}{dt}\left(\frac{dy}{dt}\right) = \frac{d^2y}{dt^2}. \quad \ldots \quad \ldots (5)$$

A body is projected with velocity *u* at an angle α to the horizon; to find its path. (Here, g is supposed to be constant and resistance of air is neglected).

Let us take the starting point as origin and the axis of co-oridinates horizontal and vertical respectively.

Let (x, y) be the co-ordinates of the position P of the body at any time t.

Since there is no force, and hence no acceleration in the horizontal direction and since the vertical accleration is always the same, being equal to g downwards, we have here the equations of motion.

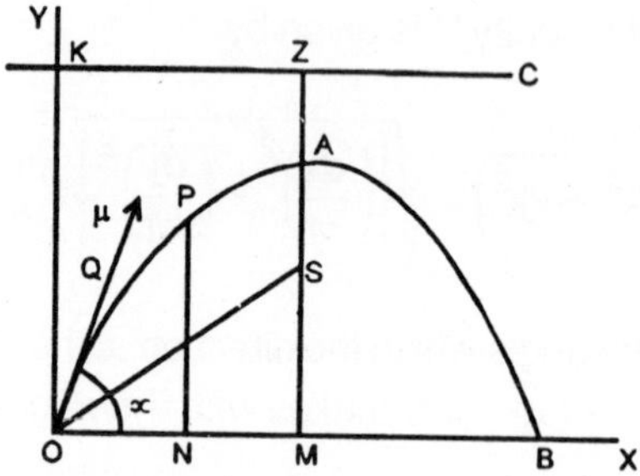

Fig. 10.1:

$$\frac{d^2x}{dt^2} = 0, \qquad \text{.... (1)}$$

$$\frac{d^2y}{dt} - g. \qquad \text{.... (2)}$$

Integrating, $\frac{dx}{dt} = A$ ... (3) $\frac{dy}{dt} = -gt + B.$ ... (4)

Since, initially, *i.e.*, at time $t = 0$, horizontal and vertical components of velocity are $u \cos \alpha$, $u \sin \alpha$ it follows from (3) and (4),

$A = u \cos \alpha$, $B = u \sin \alpha$ ... ... (5)

Thus, (3) and (4) become

$\frac{dx}{dt} = u \cos \alpha$ ... (6) $\frac{dy}{dt}$ $u \sin \alpha - gt.$ ... (7)

Integrating again (6) and (7),

$x = u\cos\alpha.t + C$ ... ... (8)

$y = u\sin\alpha.t - \frac{1}{2}gt^2 + D.$ ... ... (9)

Since, when $t = 0$, $x = 0$ we get from above, $C = 0$, $D = 0$.

$\therefore$ (6) and (7) become

$x = u\cos\alpha.t$ ... ... (10)

$y = u\sin\alpha.t - \frac{1}{2}gt^2$ ... ... (11)

From (10), $t = \dfrac{x}{u\cos\alpha}$.

Substituting this value of t in (11), we get

$$y = x\tan\alpha - \frac{1}{2}g\frac{x^2}{u^2\cos^2\alpha} \quad \text{... ... (12)}$$

which is the equation of the path of the body, and the equation being of the form $y = Ax + Bx^2$ (A and B being constants) it is the equation to a parabola.

Hence, the path of the projectile is a parabola.

Deductions:

(1). The equation (12) can be written as

$$x^2 - \frac{2u^2\sin\alpha\cos\alpha}{g}x = -\frac{2u^2\cos^2\alpha}{g}y$$

or,

$$\left(x - \frac{u^2\sin\alpha\cos\alpha}{g}\right)^2 = -\frac{2u^2\cos^2\alpha}{g}\left(y - \frac{u^2\sin^2\alpha}{2g}\right)$$

$$\text{or, } \left(x - \frac{u^2 \sin\alpha 2\alpha}{2g}\right)^2 = -\frac{2u^2 \cos^2\alpha}{g}\left(y - \frac{u^2 \sin^2\alpha}{2g}\right) \qquad \text{... (13)}$$

Transferring the origin to the point $\left(\frac{u^2 \sin 2\alpha}{2g}, \frac{u^2 \sin^2\alpha}{2g}\right)$, the equation becomes $X^2 = -\frac{2u^2 \cos^2\alpha}{g} Y.$ ... ... (14)

Hence, *the path is a parabola*, whose axis is verical and drawn downwards and whose

$$\text{Vertex is } \left(\frac{u^2 \sin 2\alpha}{2g}, \frac{u^2 \sin^2\alpha}{2g}\right) \qquad \text{... ... (15)}$$

$$\text{and latus rectum is } \frac{2u^2 - \text{Cos}^2\alpha}{g} \qquad \text{... ... (16)}$$

(2) Putting $y = 0$ in (12), we get either $x = 0$ (i.e.) when the body is at 0), or $x = \frac{u^2 \sin 2\alpha}{g}$ (i.e., when the particle is at B, hence $OB = \frac{u^2 \sin 2\alpha}{g}$). *OB* is called the horizontal range.

$$\therefore \textbf{Horizontal Range} = \frac{u^2 \sin^2\alpha}{g} \qquad \text{... ... (17)}$$

This horizontal range is maximum when $\sin 2\alpha = 1$, *i.e.*, $\alpha = ¼\pi$ or 45°.

$$\therefore \textbf{Maximum Horizontal Range} = \frac{u^2}{g}. \qquad \text{... ... (18)}$$

Thus, for a given velocity of projection the horizontal range is maximum when the angle of projection is 45°.

(3) Putting y = 0 in (11), we get either t = 0 (i.e., when the body is at 0 at the time of start), or $t = \frac{2u\sin\alpha}{g}$ (*i.e.*, when it is at *B*). The time from *0* to *B* is called time of flight. If *T* be the time of flight then

$$T = \frac{2u\sin\alpha}{g} \quad \ldots \quad \ldots (19)$$

(4). Putting $\frac{dy}{dt} = 0$ in (7), we get $t = \frac{u\sin\alpha}{g}$ = and substituting the value of t in (11), we have $y = \frac{u^2\sin^2\alpha}{g}$. Thus, if *T'* denotes the time to the greatest height and H denotes the greatest height attained, then

$$T' = \frac{u\sin\alpha}{g} \quad \ldots \quad \ldots (20)$$

$$H = \frac{u^2 \operatorname{Sin}^2\alpha}{2g} \quad \ldots \quad \ldots (21)$$

(5) From figure, height of the directrix

= *MZ* = *MA* + *AZ*

= height of the vertex + ¼ (latus rectum)

$$= \frac{u^2\sin^2\alpha}{2g} + \frac{1}{4}\frac{2u^2\cos^2\alpha}{g} = \frac{u^2}{2g}.$$

$$\therefore \textbf{Height of the directrix} = \frac{u^2}{2g} \quad \ldots \quad \ldots (22)$$

In the figure given just above draw OK perpendicular on the directrix and join $\overline{OS}$. Then we know $OS = OK$ (from the property of parabola).

$$\therefore \; OS = \frac{u^2}{2g} \qquad \ldots \qquad \ldots (23)$$

(6) Also, height of the focus = MS

= height of the vertex - ¼(latus rectum)

$$= \frac{u^2 \sin^2 \alpha}{2g} - \frac{1}{4}\frac{2u^2 \cos^2 \alpha}{g} = -\frac{u^2 \cos^2 2\alpha}{2g}.$$

Since OM = ½ (horizontal range) = $\frac{u^2 \sin 2\alpha}{2g}$ therefore,

co-ordinate of the focus S are

$$\left(\frac{u^2 \sin 2\alpha}{2g} - \frac{u^2 \cos 2\alpha}{2g}\right). \qquad \ldots \qquad \ldots (24)$$

Thus, the focus lies above, on or below the horizontal line $\overleftrightarrow{OX}$ according as $-u^2 \cos 2\alpha/2g$ is positive, zero, or negative, *i.e.*, according as $\alpha >= \text{or} < \frac{1}{4}\pi$.

Again, let a body be projected horizontally from A with velocity u.

With A as origin, take a pair of rectangular axis AX and AY, AX horizontal and AY vertical.

Let $P(x, y)$ be the position of the body on the path at any time t. Since there is no acceleration in the horizontal direction,

$$\therefore \; \frac{d^2x}{dt^2} = 0$$

$$\therefore \text{ integrating, } \frac{dx}{dt} = C \quad \ldots \quad \ldots (1)$$

Since initially, *i.e.*, when the body is at *A* the horizontal velocity is *u*,

$$\therefore \text{ when } t = 0, \quad \frac{dx}{dt} = 0, \quad \therefore \text{ from (1), } C = u$$

$$\therefore \text{ (1) becomes } \frac{dx}{dt} = u \quad \ldots (2)$$

Again integrating (2), $x = ut + D$ ... (3)

Since initially *i.e.* when the body is at *A*,

i.e., when $t = 0$, $x = 0$ $\therefore$ from (3), $D = 0$

$\therefore$ $x = ut$ ... ... (4)

Since the acceleration in the vertically downward direction is g.

$$\therefore \frac{d^2y}{dt^2} = -g$$

$$\therefore \text{ integrating, } \frac{dy}{dt} = -gt + E \quad \ldots (5)$$

Since initially there is no velocity in the vertical direction,

$$\therefore \text{ when } t = 0, \frac{dy}{dt} = 0 \therefore E = 0$$

$$\therefore \text{ (5) becomes } \frac{dy}{dt} = -gt \quad \ldots (6)$$

Integrating, $y = -\frac{1}{2}gt^2 + F$

when $t = 0$, $y = 0$ $\therefore$ $F = 0$

$\therefore$ $y = -\frac{1}{2}gt^2$. ... (7)

Eliminating *t* between (4) and (7)

$$y = -\tfrac{1}{2}g.\frac{x^2}{u^2}$$

$$\text{i.e., } x^2 = \frac{2u^2}{g}\ y. \quad \dots \quad \dots (8)$$

We know that it represents a parabola whose vertex is at the origin *i.e.,*at *A*, is *AM* and whose latus rectum is $2u^2/g$ and whose concavity is towards the negative side of *y*-axis, *i.e.*, whose axis is along *AM.*

Equations of Motion

It is given in Newton's Second Law of Motion that if a force *P* acting on a body of mass m in a given direction produces acceleration *f* (which is in the same direction) then

$$P = mf.$$

From this we have analytically this following equations of motion of a body in a given direction.

$$P = \frac{d^2x}{dt^2} = m\frac{dv}{dt} = mv\frac{dv}{dx}$$

x being the distance of the body from a fixed origin in the given direction at time t and v the velocity at that instant.

The following equations are for motion in a plane

$$X = m\frac{d}{dt^2}, Y = m\frac{d^2y}{dt^2},$$

Where *X* and *Y* are forces acting on the mass *m* along the *x*-axis and *y*-axis respectvely.

Chapter-11

IMPORTANT FORMULA FOR PRACTICAL APPLICATION

1. Resulatant of two Velocities.

$$w = \sqrt{u^2 + v^2 + 2uv\cos\alpha}\cdot$$

$$\theta = \tan^{-1} \frac{v\sin\alpha}{u + v\cos\alpha}$$

2. Rectilinear Motion with Uniform Acceleration.

$$v = u + ft.$$

$$s = ut + ½ft^2$$

$$v^2 = u^2 + 2fs.$$

Space described in the tth sec.

$$st = u + ½\, f\,(2t - 1).$$

Average velocity = $½\,(u + v)$

$$= u + ½ft.$$

3. Rectilinear Motion.

$$\text{Velocity } v = \frac{dx}{dt} : \text{acceleration } f = \frac{dv}{dt} = \frac{d^2x}{dt} = v\frac{dv}{dx}$$

4. Vertical Motion

g =32 ft. per sec^2, or, 981 cms. per sec^2.

For upward motion

Greatest height = $\dfrac{u^2}{2g}$

Time to the greatest height = $\dfrac{u}{g}$

Total time of flight = $\dfrac{2u}{g}$.

For downward motion

$$v = \sqrt{2gh}, \quad t = \sqrt{\frac{2h}{g}}.$$

5. Motion on a plane

$$v_x = \frac{dx}{dt}, v_y = \frac{dy}{dt}, f_x = \frac{d^2x}{dt^2} = \frac{dv_x}{dt}, f_y = \frac{d^2y}{dt^2} = \frac{dv_y}{dt}.$$

6. Laws of Motion

$P = mf$

$W = mg$

$$P = m\frac{d^2x}{dt^2} = m\frac{dv}{dt} = mv\frac{dv}{dx}$$

$$X = m\frac{d^2x}{dt^2}, \qquad Y = m\frac{d^2y}{dt^2}.$$

7. Projectiles.

Greatest height = $\dfrac{u^2 \sin^2 \alpha}{2g}$

Time to the greatest height = $\dfrac{u \sin \alpha}{g}$

Total time of flight = $\frac{2u\sin\alpha}{g}$

Horizontal Range = $\frac{u^2 \sin 2\alpha}{g}$

Maximum Horizontal Range = $\frac{u^2}{g}$

Latus rectum of the path = $\frac{2u^2 \cos^2 \alpha}{g}$

Velocity at height h, $v^2 = u^2 - 2gh$

Height of the directrix = $\frac{u^2}{2g}$.

Distance of the focus from the point of projection = height of the directrix.

Range up an inclined plane (inclination β)

$$R = \frac{u^2}{g\cos^2\beta}[\sin(2\alpha - \beta) - \sin\beta]$$

Max. oblique range up = $\frac{u^2}{g(1+\sin\beta)}$

Range down an inclined plane (inclination β)

$$R = R = \frac{u^2}{g\cos^2\beta}[\sin(2\alpha - \beta) + \sin\beta]$$

Max. oblique range down = $\frac{u^2}{g(1-\sin\beta)}$

Analytical equation of the path of a projectile

$$y = x \tan \alpha - \tfrac{1}{2} g \frac{x^2}{u^2 \cos^2 \alpha}$$

Co-ordinates of the vertex $A \left(\frac{u^2 \sin^2 2\alpha}{2g}, \frac{u^2 \sin^2 \alpha}{2g} \right)$

Co-ordinates of the focus $S \left(\frac{u^2 \sin^2 2\alpha}{2g}, -\frac{u^2 \cos 2\alpha}{2g} \right)$

8. Simple Harmonic Motion.

Accleration = $\frac{d^2x}{dt^2} = -\mu x$ (x = distance from the centre of oscillation $\mu > 0$)

Velocity = $\frac{dx}{dt} = \pm\sqrt{\mu(a^2 - x^2)}$

(a being the amplitude)

$x = a \cos \left(\sqrt{\mu t + \varepsilon}\right)$

Period = T = $2\pi / \sqrt{\mu}$.

Chapter-12

SUMMARY OF APPLICATIONS OF MECHANICAL LAWS AND PRINCIPLES TO HUMAN MOTION

A review is useful in mechanical anaysis of motion, to solve problems and perform the laboratory experiences.

Displacement, velocity, and force may be represented by vectors, that is arrows indicating of vectors the graphic polygon method of solution is recommended as the most simple. For instance, consider the following four vectors based on compass points representing displacement.

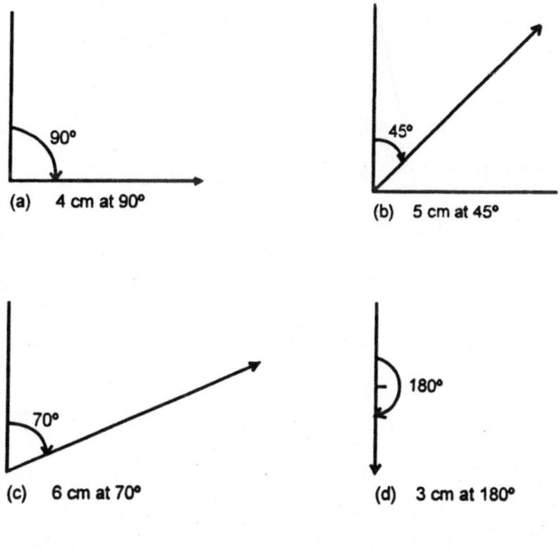

Fig. 12.1:

To sum them and find the resultant, the arrow may be drawn as represented below, with R representing the summed displacement.

An example using triangulation could be made of an athlete performing the long jump. Suppose the horizontal running velocity is 8 meters per second and the vertical velocity imparted at takeoff is 6 meters per second; the resultant velocity is the graphic sum of the athletics.

$$R = \sqrt{(8)^2 + (6)^2} = \sqrt{100} = 10 \text{ meters per second}$$

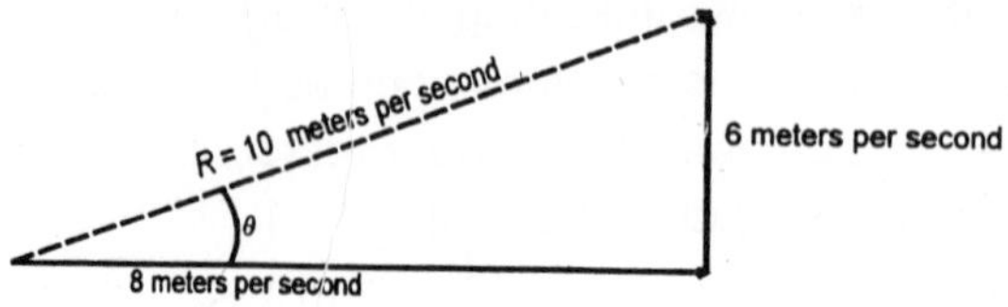

Fig. 12.2:

In this instance, the angle of projection (θ) is defined by the following trigonometric relationship. The sin θ = $\frac{\text{opposite side}}{\text{hypotenuse}}$ or $\frac{6\ m/s}{10\ m/s} = 0.6$ (which is the arc sine). Therefore θ = 20.43° with the horizontal. The same answer is achieved by a second trigonometric relationship as follows. The tan θ = $\frac{\text{opposite side}}{\text{adjacent side}}$ or $\frac{6\ m/s}{10\ m/s} = 0.336$ (which is the arc tangent). Therefore q = 20.43° with horizontal. Each trigonometric relationship seeks and achieves the same answer. (To test these values see the trigonometric table or use a pocket calculator).

Because problems involving triangular configurations will be encountered in many situations, pertinent trigonometric functions and their definitions are given here. Notice the following statement about a right triangle.

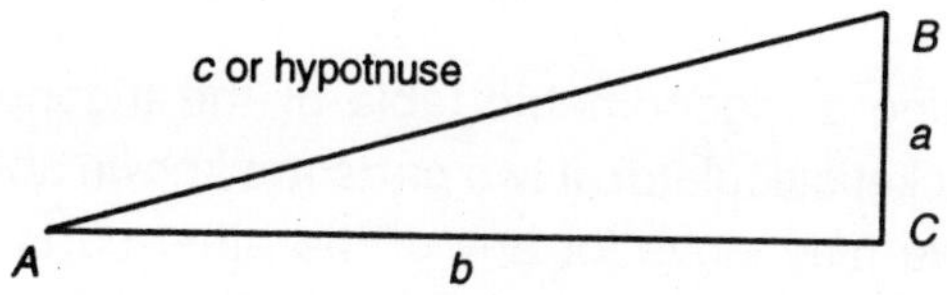

Fig. 12.3:

In the given right triangle *ABC*

1. The sine of an angle $\frac{\text{length of side opposite that angle}}{\text{length of hypotenuse}}$

$$(\sin A = \frac{a}{c})$$

2. The cosine of an angle =

$$\frac{\text{length of side adjacent to that angle}}{\text{length of hypotenuse}}$$

$$(\cos A = \frac{b}{c})$$

3. The tangent of an angle =

$$\frac{\text{length of side opposite that angle}}{\text{length of side adjacent to that angle}}.$$

$$\tan A = \frac{a}{b})$$

4. The cotangent of an angle

$$= \frac{\textit{length of side adjacent to that angle}}{\textit{length of side opposite that angle}}$$

$$(\cot A = \frac{b}{a})$$

Thus, by using a trigonometric table or the trigonometric function of a pocket calculator, if two parts are known about the right triangle, we may slove for any of the other parts. There are, of course, six parts: three angles (*A*, *B*, *C*) and three sides (*a*, *b*, *c*). If the A or B is known, the other is arrived at by simply subtracting the known angle from 90°.C always is equal to 90°, and the sum of the three angles in any triangle is equal to 180°. in dealing with force or velocity vectors, the same system is used.

The following formulas are commonly used when dealing with linear kinematics, angular kinematics, kinetic, and equilibrium.

A. Linear Kinematics (Linear Motion)

1. $v = \frac{d}{t}$ where v is velocity; d is displacement or distance; and *t* is time

2. $\bar{v} = \frac{u + v}{2}$ where $\bar{v}$ is average velocity; *u* is initial veloctiy; and *v* is final velocity

3. $\bar{a} = \frac{v - u}{t}$ where $\bar{a}$ is average acceleration

4. $v = u + at$ where *v* is final velocity when acceleration is uniform (derived from $a = \frac{v - u}{2}$)

5. $s = ut + ½at^2$ where s is vertical distance traveled when acceleration is constant, as in the case of gravity (gravity constant $g = 9.8$ m/s² or 32.2 ft/s²)

6. $s = \frac{1}{2} gt^2$ when initial velocity is zero

In dealing with projectiles, freely falling bodies, and gravity we use the following.

7. $v_y^2 = +u_y^2 + 2gs$ where v_y is vertical takeoff velocity; u_y is initial vertical velocity at top of flight

8. $V_y^2 = +2gs$ when initial velocity is zero

If the angle of takeoff and landing (θ) is other than vertical (90°), then the following projectile formulas, as derived from the above, apply.

9. $t = \frac{u \sin\theta}{g}$ where t is time required for an object to reach its highest point

or

10. $t = \sqrt{2s/g}$ where t is time to fall from highest point

11. $T = \frac{2u \sin\theta}{g}$ where T is time object is in flight

12. $s = \frac{(u \sin\theta)^2}{g}$ where s is vertical height the object attains

13. $R = \frac{u^2 \sin 2\theta}{g}$ where R is horizontal distance the object covers from takeoff to landing at the same level as takeoff

14. $$R = \frac{u^2 \sin\theta \cos\theta + u \cos\theta \sqrt{(u \sin\theta)^2 + 2gh}}{g}$$

where R is range of a projectile when height of takeoff is different from height of landing; h is height difference between takeoff and landing

15. $v_x = v_\theta \cos\theta$ where v_x is horizontal velocity; v_θ is velocity in the direction of projection

16. $v_y = v_\theta \sin\theta$ where, v_y is vertical velocity

17. $v_x = v_y \tan\theta$

18. $v_y = v_x \tan\theta$

19. $v_\theta = v_x/\cos\theta = v_y/\sin\theta = \sqrt{v_x^2 + v_y^2}$

B. Angular Kinematics (Angular Motion)

20. $\omega = \frac{\theta}{t}$ where ω (omega) is angular velocity

21. $\bar{\alpha} = \frac{\omega_f - \omega_t}{t}$ where $\bar{\alpha}$ (alpha) is average angular acceleration

22. (a) $v_T = \omega r$ where v_T is tangential velocity (linear, curvilinear, or circular velocity at a point
(b) $\omega_1 r_1 = \omega_2 r_2$

23. $V_T = \dfrac{\omega r}{57.3}$ converts ω from degrees to radians, hence to linear units

24. $C = 2\pi r$ where C is circumference of circle and π is 22/7 or 3.1416; therefore, 1 radian equals length of the radius on circumference

25. 1 rev = 360°

26. 360° = 6.28 radians

27. 1 radian = 57.3°

C. Kinetics (Force)

28. $Wt = mg$ where Wt is weight; m is mass ; g is gravity

29. $m = \dfrac{Wt}{g}$

30. $T = Fr$ where T is torque; F is force applied perpendicular ot the radius (usually expressed in meter-newtons)

31. $F \times FA = R \times RA$ when force act perpendicular (at 90°) to the radius of rotation. where F is force; FA is length of force arm; R is resistance; RA is length of resistance arm

If forces act at other than 90° to the arms

32. $F \times FMA = R \times RMA$ where MA is moment arm for force or resistance (or this must be calculated using trigonometry)

33. $M.A. = \frac{R}{F}$ where *M.A.* is mechanical advantage

and

34. $M.A. = \frac{FA}{RA}$

35. $W = Fd$ where W is work; F is force; d is distance (usually expressed in meter-newtons or foot-pounds)

36. $P = \frac{W}{t}$ where P is power, the rate at which work is accomplished

or $P = \frac{Fd}{t}$

37. $P.E. = mgh = W \times h$ where *P.E.* is potential energy; m is mass; g is gravity; h is height of centre of gravity of mass above ground level; also mg equal Wt, since m is $\frac{Wt}{g}$

38. $P.E. = K.E.$ where *K.E.* is kinetic energy

As long as a body moves it has kinetic energy in proportion to its mass and velocity. Thus

39. $K.E. = ½mv^2$

Because of the principle of conservation of momentum

40. $W = K.E. = Fd$ or work done is equal to kinetic energy acquired

Therefore,

41. $Fd = ½mv^2$

D. Kinematics and Kinetics (Motion and Force)

Newton's Laws of Motion Newton's second law, the law of acceleration, states that the force developed is in proportions to the product of the mass and acceleration of an object. Mass is usually expressed in kilograms and acceleration in meters per second squared (m/s^2). Force is expressed in newtons.

42. $F = ma$ where F is force;m is mass ; a is acceleration. From that equation we derive the impulse equation; if we set

$$a = \frac{v-u}{t}, \text{ then } F = \frac{m(v-u)}{t} \text{ and}$$

43. $Ft = mv - mu$

44. Impulse = Ft or, Impulse is the product of force and the time over which the force acts

45. $M = mv$ where M is momentum, m is mass; v is velocity

Newton's third law of motion, or reaction and conservation of momentum in a system, states that the momentum lost by a striking object is gained by the object which is struck, or

46. $M_1 v_1 - M_1 u_1 = M_2 v_2 - M_2 u_2$

Rotational Equivalents

Newton's first principle, the law of inertia, is expressed by the formula which says that the moment of inertia about an axis is the sum of the mass of particles multiplied by the square of the distance of each particle from the axis. It expresses the force required to start or stop motion and is written.

47. $I = mr^2$

48. $A.M. = I\omega$ where $A.M.$ is angular momentum; I is intertia; ω is angular velocity ($\omega = v_r/r$)

and

49. $A.M. = mrv_r$ where m is object's mass; r is radius of rotation at object's centre of gravity; v_r is tangential velocity at that point (derived from $A.M. = mr^2 [v_r/r] = mrv_r$)

The centripetal force (Fc) is centre-seeking and is equal in magnitude and opposite in direction to the centrifugal. This is the force required to hold an object in a curved path and is expressed in the formula

50. $Fc = \frac{mvr^2}{r}$ where m is object's mass; v_T^2 is object's tangential velocity squared; r is radius measured from object's centre of rotation to centre of gravity

Forces Modifying Motion

The coefficient of friction (μ) is the quotient of the force of pull (P) needed to overcome the friction caused by weight of the object holding the two surfaces together. The value is usually less than 1. It is expressed as

51. $\mu = \frac{P}{Wt}$

The coefficient of friction is the arc tangent and determines the amount of lean possible by a player or

52. $\tan\theta = \frac{P}{Wt} = \mu$ where θ is an angle of lean with the vertical

53. $\varepsilon = \sqrt{\frac{h_b}{h_d}}$ where ε is coefficient of elasticity or

resolution of an object; h_b is height of bounce; h_d is height of drop

The neutral coefficient of restitution of two moving objects can be determined by a formula based on the conservation of momentum.

54. $\varepsilon = \dfrac{v_2 - v_1}{u_1 - u_2}$

E. Force and Equilibrium

The formula for determining the location of the centre of gravity of an individual in a lying position on a reaction board used with a weighing scale is expressed as

55. $X = \dfrac{Fl}{Wt}$ where X is distance above soles of the feet; F is scale reading of partial weight of subject and board minus scale reading of partial weight of board; l is length of reaction board between knife-edges; and Wt is weight of subject

56. $\%h = \dfrac{x}{h} \times 100$ is the location of centre of gravity (as measured from soles of feet) where h is total height of individual

The following prefixes indicate the most commonly used factors.

micro	0.000001	10^{-6}
milli	0.001	10^{-3}
centi	0.01	10^{-2}
kilo	1,000	10^{3}
mega	1,000,000	10^{6}

Conversion Table

	To change from	To	Multiply by or to reverse divide by
Acceleration	Meters per second^2	Feet per second2	3.28
	Meter per second^2	Miles per hour per second	2.237
Angle	Radians	Degrees	57.296
	Radians	Revolutions	0.159
	Revolutions	Degrees	360
	Revolutions	Radians	6.28 or
Distance	Meters	Feet	3.28
	Meters	Yards	1.094
	Kilometers	Miles	0.621
	Centimeters	Inches	0.3937
Energy	Meter-newton (joules)	Foot-pounds	0.738
	Joules	Kilocalories	0.00023
	Foot-pounds	Joules (meter-newtons)	1.358
Force	Newtons	Founds	0.225
Impulse	Newton-seconds	Pounds-seconds	0.225
Mass	Kilogram	Slugs	0.0685
Moment of inertia	Kilogram-meters^2	Slug-feet^2	0.738
	Kilogram-meters^2	Pound-feet^2	23.7
	Slug-feet^2	Pound-feet^2	32.174
Momentum	kilogram-meters per second	Slug per feet per second	0.225
	Kilogram-meter per second	Pound-feet per second	7.227
	Slug-feet per second	Pound-feet per second	32.174
Power	Kilocalories per second	Watts	4186
	Kilocalories per second	Horsepower	0.00156
	Kilocalories	Foot-pound per second	0.858
	Horsepower	Foot-pounds per second	550.
	Horsepower	Watts	745.2
	Foot-pounds per second	Watts	1.356
Pressure	Kilograms per centimeter^2	Pounds per inch^2	14.22
	Kilograms per meter^2	Pounds per inch^2	0.00162
	Newtons per centimeter^2	Pounds per inch^2	0.0145
	Dynes per centrimeter^2	Pounds per inch^2	1.45
Torque	Meter-newtons	Pounds-feet	0.737
Velocity	Meters per second	Feet per Second	3.28
	Kilometers per hour	Mies per hour	0.621
	Miles per hour	Feet per second	1.467
	Meters per second	Miles per hour	2.237

Chapter-13

PRACTICALS FOR ADVANCED STUDENTS

13.1 DETERMINATION OF C.G. BY ANALYTICAL METHOD

C.G. is taken as reference point for any motion. Two methods are used to determine it:

1. Experimental Method
2. Analytical Method

   A – Joint Point Method

   B – Main Method

13.2 (A) DETERMINATION OF C.G. BY JOINT POINT METHOD

A photograph of an athlete is used for the purpose of analytical method: Based upon the principles of moment, certain fixed relationships of a body part are derived with whole body in normal conditions. (mentioned on performa). Refer the photograph below;

Moment = Equilant of force in rotation

= force × distance.

On 'X' axis:

$M = 100 \times d$

$d \times 100 = 625.39$

$d = \frac{625.39}{100} = 6.25 = \underline{6.3 \text{ cm.}}$

On 'Y' axis

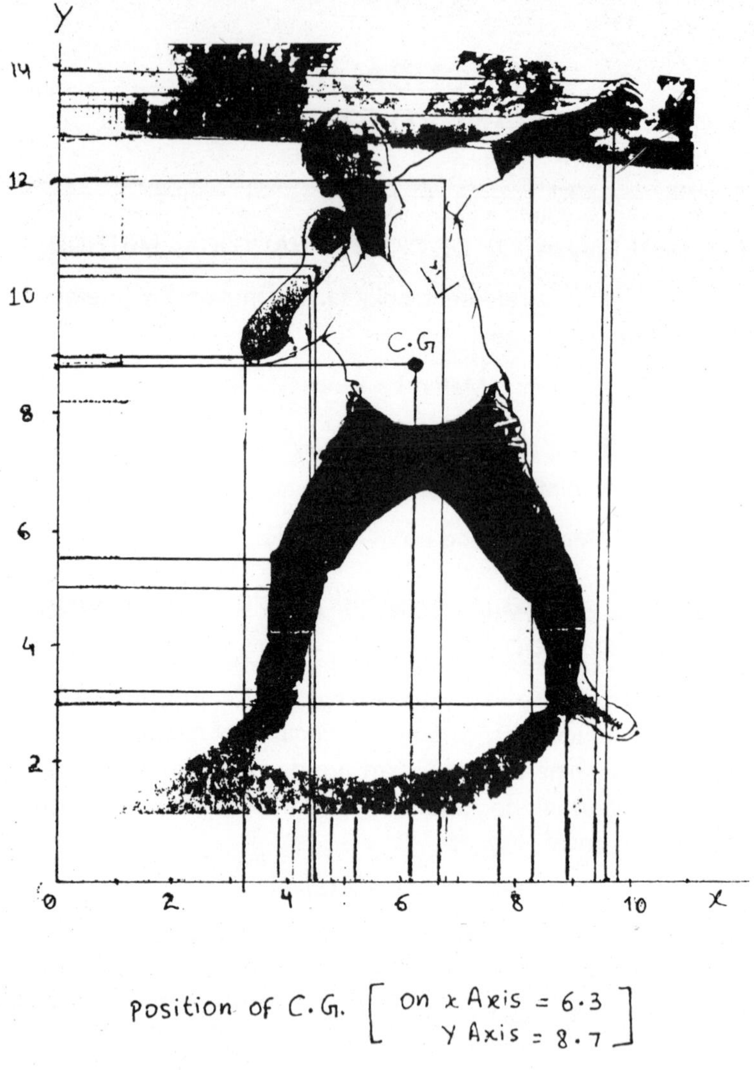

Fig. 13.1: Determination of C.G. by Joint Point Method

$$M = 100 \times d$$

$$d \times 100 = 872.22$$

$$d = \frac{872.22}{100} = 8.72 = \underline{8.7\text{ cm.}}$$

Body Part		Rel.Weight (%)	X (cm.)	Y (cm.)	XG (cm.)	YG (cm.)
Head		7.0	5.3	12.8	37.10	89.60
Shoulder Joint	(R)	13.6	4.9	10.4	66.64	141.44
	(L)		6.9	12.1	91.12	164.56
Elbow Joint	(R)	2.6	3.3	9.0	8.58	23.40
	(L)		8.4	12.7	21.84	33.02
Wrist	(R)	0.8	4.5	10.6	3.60	8.48
	(L)		9.7	13.4	7.76	10.72
Hand	(R)	1.0	4.6	10.7	4.60	10.10
	(L)		9.8	13.5	9.80	13.50
Hip Joint	(R)	16.2	5.3	8.2	85.86	132.84
	(L)		7.8	8.2	126.36	132.84
Knee Joint	(R)	8.2	4.2	5.5	34.44	45.10
	(L)		9.0	5.0	73.80	41.00
Ankle Joint	(R)	2.1	3.9	3.1	8.19	6.51
	(L)		9.0	3.1	18.90	6.51
Foot	(R)	2.0	3.9	3.0	7.80	6.00
	(L)		9.5	3.0	19.00	6.00
		100.00			625.39	871.62

$X = 6.3$ $\quad$ $Y = 8.7$

$M = 100 \times d$ $\quad$ $M = 100 \times d$

$d = \dfrac{625.39}{100}$ $\quad$ $d = \dfrac{872.22}{100}$

$= 6.3$ $\quad$ $= 8.7$

13.3 (B) DETERMINATION OF C.G. BY MAIN POINT METHOD

Aim

To determine the C.G. of the athlete using the 'Main Point' method.

Theory

The body is divided into compartments e.g. thigh, leg, upper arm, forearm etc. The C.G. of each compartment is calculated individually depending on its length and relative weight of its proximal and distal parts. The combined C.G. of the body is calculated by summation.

Procedure

The picture of the athlete is pasted on a graph paper.

The length of each compartment is measured. This is multiplied by a factor is characterstic for that compartment depending on the position from the proximal end and is marked in that compartment. The abscissa and ordinate of this marked point are located these are multiplied by the relative weights of each body compartment.

Hence, the C.G. is found out.

The C.G. found out to be (13.65, 14.05) is hence plotted on the photograph.

***Note*:**

In case of C.G. of individual compartments, fraction can also be subsitituted in place of decimals. In such case the function = ratio between proximal and distal position from C.G.

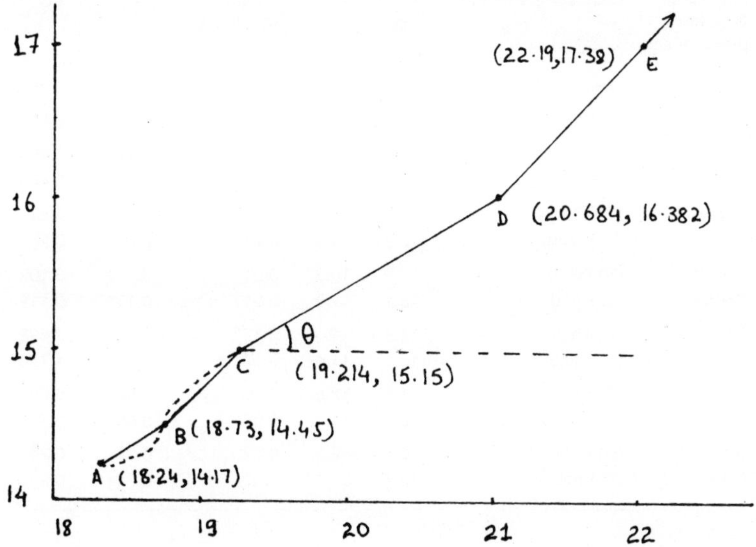
17
16
15
14
18
19
20
21
22
(22.19,17.38)
E
D (20.684, 16.382)
θ
C
(19.214, 15.15)
B (18.73, 14.45)
A (18.24,14.17)

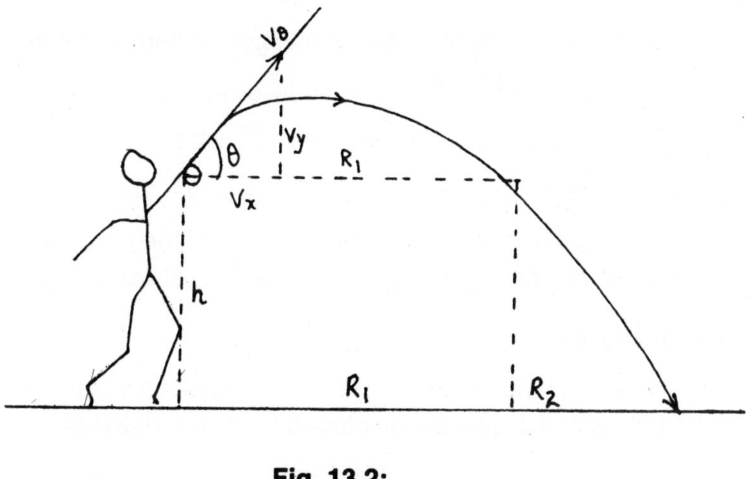
Vθ
Vy
θ
R1
Vx
h
R1
R2

Fig. 13.2:

Position of C.G. from proximal end (%)	Body compartments	X	Y	Relative weight (G)	XG	YG
Centre	Head	16.1	13.8	0.07	1.127	0.966
0.44	Trunk	14.0	14.9	0.43	6.03	6.40
0.47	(R) U. Arm	14.3	13.6	0.03	0.429	0.408
0.47	(L) U. Arm	15.3	14.3	0.03	0.459	0.429
0.42	(R) F. Arm	14.0	12.0	0.02	0.28	0.24
0.42	(L) F. Arm	15.0	13.0	0.02	0.30	0.26
Centre	(R) Hand	13.9	10.9	0.01	0.139	0.109
Centre	(L) Hand	15.1	11.9	0.01	0.151	0.119
0.44	(R) Thigh	12.3	14.1	0.12	1.476	1.692
0.44	(L) Thigh	13.1	11.4	0.12	1.572	1.728
0.42	(R) L. Leg	11.5	12.6	0.05	0.575	0.63
0.42	(L) L. Leg	13.3	11.9	0.05	0.665	0.595
0.44	(R) Foot	10.5	12.0	0.02	0.21	0.24
0.44	(L) Foot	12.7	11.74	0.02	0.254	0.234

$\epsilon x_{CG} = 13.65$ $\epsilon y_{CG} = 14.05$

13.4 DETERMINATION OF CENTRE OF GRAVITY BY THE 'JOINT POINT' METHOD WITH EXPLANATION

Theory

Centre of gravity of a body is that point where all force act.

The calculaion by this method involves a set of reference point i.e., the Head, Shoulder joint $<^{R}_{L}$, Elbow $<^{R}_{L}$, Wrist $<^{R}_{L}$, Hand $<^{R}_{L}$, Hip $<^{R}_{L}$, Knee joint $<^{R}_{L}$, Ankle $<^{R}_{L}$, Foot $<^{R}_{L}$. Each reference point is given a relative weight i.e., the percentage of the body weight e.g., (R) Elbow joint 2.6 % (of body weight).

Procedure

Now the photograph of the athlete is pasted on a graph paper and the *X* and *Y* axis are numbered from zero onwards.

The reference points are made on the photograph and from these point, perpendicular lines are drawn to the *X* and *Y* axis.

If a joint is not visible, its position is assumed (i.e. it is marked where it is thought to be.

The abscissa is marked as *X* and ordinated as *Y*.

The relative weight percent is marked as G.

The points (*X*, *Y*) on the graph each relative point are marked.

XG and *YG* is calculated for each (*X* and *Y*) $\in x_{C.G.}$ and $\in y_{C.G.}$ is then summed $\frac{\in xcg}{100}$ gives the *X* axis value of C.G. and $\frac{\in ycg}{100}$ gives the *Y* axis value of C.G. $\frac{\in x_{C.G.}}{100}$ and $\frac{\in y_{C.G.}}{100}$ are then plotted on the photograph.

Body Part		Rel.Weight (%)	X (cm.)	Y (cm.)	XG (cm.)	YG (cm.)
Head		7.0	14.2	10.8	99.4	75.6
Shoulder Joint	(R)	13.6	12.6	11.2	171.4	152.32
	(L)		13.7	12.2	186.3	165.92
Elbow Joint	(R)	2.6	11.8	9.7	30.68	25.22
	(L)		12.9	10.6	33.54	27.56
Wrist	(R)	0.8	11.9	8.3	9.52	6.64
	(L)		12.8	9.1	10.24	7.28
Hand	(R)	1.0	11.9	7.9	11.9	7.9
	(L)		13.0	8.9	13.0	8.9
Hip Joint	(R)	16.2	10.2	12.0	165.24	194.4
	(L)		10.9	12.3	176.58	199.26
Knee Joint	(R)	8.2	10.2	9.8	83.64	80.36
	(L)		11.7	10.3	95.94	84.46

(Contd...)

Body Part		Rel.Weight (%)	X (cm.)	Y (cm.)	XG (cm.)	YG (cm.)
Ankle Joint	(R)	2.1	8.5	9.5	17.85	19.95
	(L)		10.4	9.1	21.84	19.11
Foot	(R)	2.0	8.5	8.9	17.0	17.8
	(L)		10.8	8.8	21.6	17.6
		100.00			1165.67	1110.28

$X_{cG} = 11.66$ $Y_{cG} = 11.10$

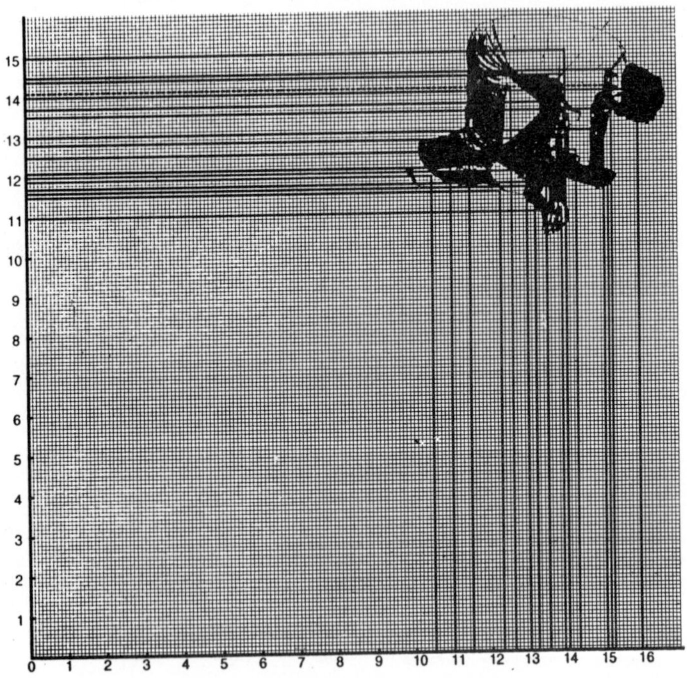

Fig. 13.3:

13.5 CINEMATOGRAPHIC METHOD

The movie photosonic camera is fixed on a tripod stand after making it horizontal with the level. The camera is set perpendicular to the plane of motion. It is placed as close as possible to the subject. Speed of the camera is usually adjusted from 50–100 frames/sec. but the provision is there for 10–1000 frames/sec. The speed is checked by means of a crystallized diode which flickers for a known interval. The number of frames on which the diode is seen is counted and speed is calculated. Alternatively a stop watch is filmed and the number of frames taken to cover 1 sec. will denote the speed.

The information regarding distance is achieved by filming a known length and converting that distance into a calibration factor (*i.e.,* actual length/length on film).

Reference points are needed as the camera may move from its position, hence, a reference point (c is fixed) is taken. Reference vertical and horizontal planes are also taken using fixed points (e.g., edge of a building).

13.6 ADJUSTMENT ON THE MOVIE CAMERA (CINEMATOGRAPHY)

1. Horizontal Position - check with spirit level
2. Focus
3. Zoom
4. Apperture Time

   16 in bright light (= 1/16 sec.)

   8 in shade (1/8 sec)
5. Apperture speed 100msec., 200 msec.
6. Film speed 50 frames/sec
7. Battery check.

13.7 DETERMINATION OF CENTRE OF GRAVITY USING CINEMATOGRAPHIC METHOD

Aim

To find out the C.G. of the athelete by a cinematograph

Theory

Field motion analyzer is the equipment by which each frame of a film can be analyzed for motion. Hence a particular frame (photograph) of the athlete can be enlarged and analyzed.

Cinematograph is the equipment on which the film is to be loaded and moved.

Procedure

The photograph of a shot putter during his action is focussed. The C.G. is calculated by the joint point method, but this time, the points of the joints are marked by the indicator of the field motion analyzer. The button on the indicator is pressed and the X and Y coordinates of that point appear on the digital monitor were recorded.

Body Part		Rel.Weight (%)	X (cm.)	Y (cm.)	XG (cm.)	YG (cm.)
Head		7.0	13.324	12.266	93.268	85.862
Shoulder Joint	(R)	13.6	12.174	11.590	165.566	157.624
	(L)		13.224	11.606	179.846	157.842
Elbow Joint	(R)	2.6	11.726	10.610	30.487	27.586
	(L)		14.216	10.676	36.961	27.757
Wrist	(R)	0.8	12.876	11.174	10.300	8.939
	(L)		15.146	10.066	12.117	8.053
Hand	(R)	1.0	12.632	11.500	12.632	11.500
	(L)		15.312	09.936	15.312	9.936

(Contd...)

Body Part		Rel.Weight (%)	X (cm.)	Y (cm.)	XG (cm.)	YG (cm.)
Hip Joint	(R)	16.2	11.392	09.524	184.550	154.289
	(L)		11.252	9.958	182.282	161.319
Knee Joint	(R)	8.2	11.774	08.030	96.547	65.846
	(L)		09.428	09.270	77.310	76.014
Ankle Joint	(R)	2.1	10.404	05.412	21.848	11.365
	(L)		07.792	09.898	16.363	20.786
Foot	(R)	2.0	10.768	05.800	21.536	11.600
	(L)		07.312	09.814	14.624	19.628
		100.00			1171.549	1016.424

$X_{c.g.} = 11.72$ $Y_{c.g.} = 10.16$

13.8 KINEMATIC ANALYSIS OF SHOT PUTTING

Aim

To determine the

(a) Path of the shot

(b) Angle of release

(c) Release velocity and

(d) Resultant by using a field motion analyzer.

Procedure

The photograph is focussed just before the release of the shot ~~put.~~

Frame speed = 62 frames/min.

$\Rightarrow$ 1 frame = 0.16 sec.

(= time between 2 frames)

The *X* and *Y* coordinates of the shot put 3 to 4 frames before release are noted down.

The *x* and *y* coordinates at the time of release are also noted down.

Also two frames after the release are focussed and the *x* and *y* co-ordinates of the shot in the air are noted down.

The motion of the shot is marked on a graph.

Path of the Shot

The coordinates of *X* and *Y* are marked on the graph for 5 points (2 before, 1 during and 2 after release)

Position	*X* coordinate	*Y* coordinate
A	18.238	14.172
B	18.730	14.450
C	19.214	15.150 ← Point of release
D	20.684	16.382
E	22.196	17.382

(In fact the complete path of the shot starting from the release to the landing can be marked but it would take more of film and time).

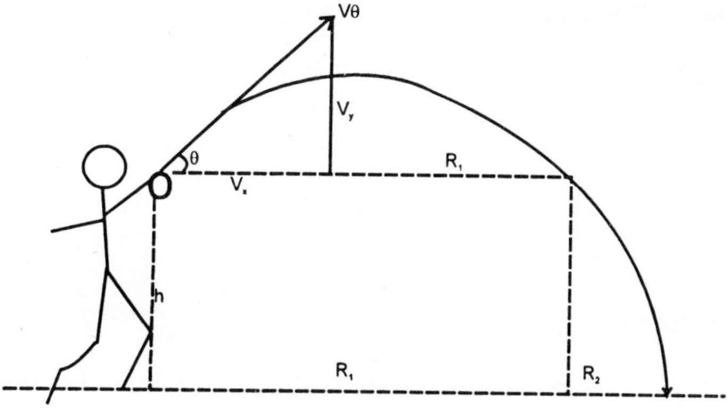

Path of the Shot Put during Release

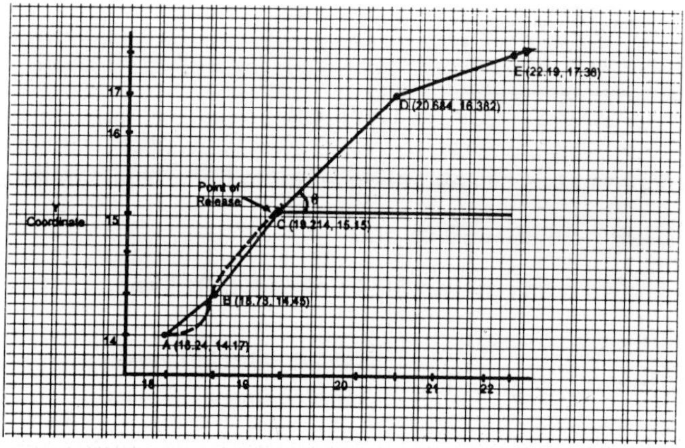

Fig. 13.4:

Calibration Factor

This is calculated by finding the ratio between the distance on the film and the actual distance. A predetermined length is measured on the film and the ratio is found out. i.e.

$$\frac{\text{Actual length}}{\text{Length of Film}} \text{ in this case} = 28.89.$$

Displacement

A point after the release is taken and the displacement is determined in X axis and Y axis respectively. From graph displacement in X axis = $X_D - X_C = 20.68 - 19.21 = 1.47$ displacement in y axis = $y_D - y_C = 16.38 - 15.15 = 1.23$.

Resultant

Applying pythagoras's theorem the resultant is found out

$$R = \sqrt{(X_D - X_C)^2 + (Y_D - Y_C)^2}$$

$$R = \sqrt{(1.47)^2 + (1.23)^2} = 1.198 \text{ cm.}$$

multiplying this by calibration factor to find actual resulatant

$= 1.198 \times 28.89 = 55.41$ cm.

$\Rightarrow$ The shot has moved 55.41 cm. (in 0.016 sec.)

$$\theta = \text{release angle} = \text{Tan}^{-1}\left(\frac{\textit{displacement in y axis}}{\textit{displacement in x axis}}\right)$$

displacement in the path of the shot =

$$\sqrt{(\textit{Displacement in x axis})^2 + (\textit{Displacement in y axis})^2}$$

= 55.41 cm.

also $V_x = \dfrac{\text{displacement in X axis}}{\textit{time}}$ = 22.06 m/sec.

also V_y = Displacement in y axis/time = 26.54 m/sec.

Angle of Release

$$\tan\theta = \left(\frac{Y_D - YC}{X_D - X_C}\right)$$

$$\Rightarrow \tan\theta = \frac{1.23}{1.47} = 0.837$$

$$\Rightarrow \theta = \tan^{-1} 0.838 = 39.96°$$

The shot was thrown at an angle of 39.96° at release.

Release Velocity

Is to be computed in 2 directions

$$\frac{\Delta D}{\Delta T} = V_x = \frac{\textit{displacement in X direction}}{\textit{Time}}$$

$$= \frac{1.232 \times 28.29}{0.016} = 2178.33 \text{ cm./sec.}$$

= 21.8 m./sec.

$$V_y = \frac{\textit{displacement in y axis}}{\textit{Time}}$$

$$= \frac{1.47 \times 28.89}{0.016}$$

= 2654.26 cm./sec.

= 26.54 m./sec.

Other parameter can be found out

$$\text{actual velocity} = \frac{\text{Resultant displacement}}{\text{Time}} = \frac{55.41}{0.016}$$

= 34.63 m/sec.

13.9 DYNAMOMETER

Transducers are used to measure the force, two types of transducers are available;

Strain Gauges

Strain changes the length of the wire hence its electrical resistance also changes. If certain current is passed through wire if resistance (R) increases then voltage (v) decreases and reversely if resistance decreases then the voltage increases. (R↑ V↓ if R↓, V↑).

On a recording device voltage is recorded and that is measure of the strain, strain α change in length of wire α *R* α *V* change in voltage α signal display.

Quartz

When strain is applied a quartz crystal due to its special property, current is generated v of low value. This is amplified and recorded.

Platform

Records force in 6 direction vertical, horizontal, depth

x, y, z (depends on made)

Standing jump = force recorded in the vertical direction.

movements for Δ running, walking, jumping.

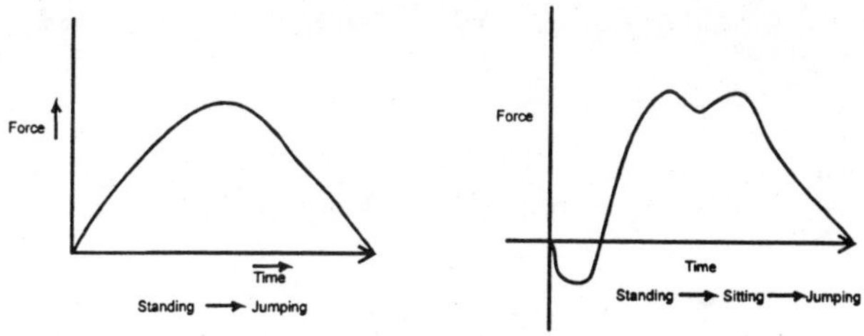

Fig. 13.5:

Counteraction

When a movement takes place, force is requared to that direction i.e., reverse to the movement is required.

Impulse = $F \times T$

= area under time curve

= count number of small squares (m.m.)

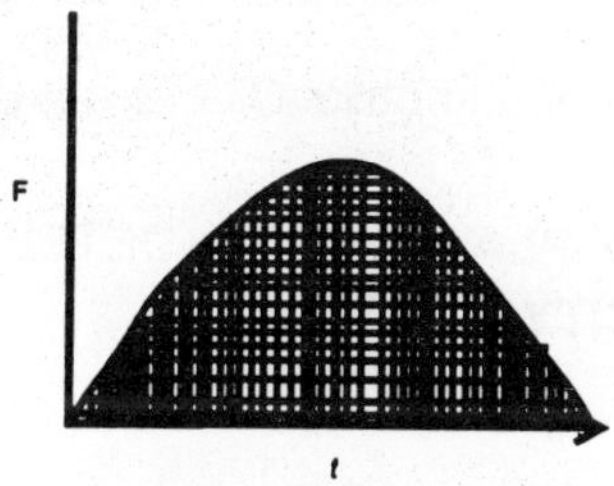

Fig. 13.6:

Fixed F = 1000 N. sec. t = 10 cm. sec.

speed of moving paper = 10 cm./sec. of the pen recorder

⇒ 1 cm. on X axis ⇒ 0.1 sec.

Force 1 cm. = 1000 Newtons

∴ 1 cm^2 = 100 Newton/sec.

Number of smalll squares × 100 Newton sec. = Impulse

$F \times T = m$ (mass) x v (velocity)

$$\Rightarrow v = \frac{Ft}{m} \text{ = velocity of centre of gravity}$$

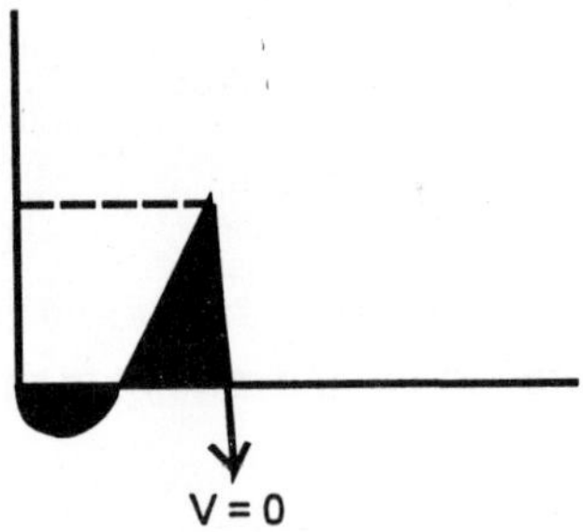

Fig. 13.7:

$v = \frac{d}{t} \Rightarrow$d = vt and the height can be calculate actually we are calculating how much body's C.G. is going up.

13.10 USE FORCE PLATFORM TO DETERMINE IMPULSE AND VELOCITY

Aim

Calculation of impulse and velocity during takeoff in a standing vertical jump.

Apparatus

- Force platform
- Amplifier + Dyanamometer
- Recorder
- Height stand
- Weighing machine

Steps

Weight is recorded.

Height of the height stand is fixed at above the subjects height.

A jump is taken from above the platform trying from above the platform trying to go as high as you can. The jump is taken as standing → sitting → jumping. The recording achieved on the graph paper is analysed.

Theory

Impulse = Force × time (area under the curve)

= mass × velocity

Force recordings are available on the y axis with the caliberation of 1 cm. = 500 Newton.

Force recordings are available on the 'X' axis with the caliberation of 1 cm. = 0.1 sec.

Transducers are present in the dynamometer which here the property of generation of electric current when strain is applied. In this dyanamometer the transducer is in the form of quartz.

The force platform records force in 3 direcrtions i.e., vertical, horizontal and depth. In a vertical jump only the vertical direction is needed.

As initially the subject goes down hence resulting in a motion with the gravity a negative force is being applied and a correction for this has to be deducted from the area under the curve.

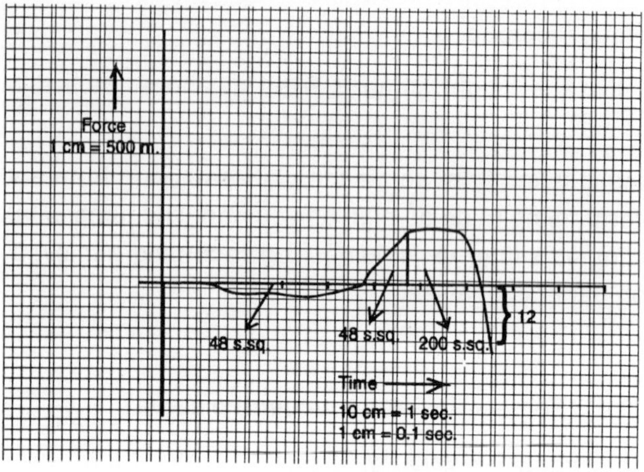

Fig. 13.8:

Calculations

(A) Area under the curve after correction = 200 small squares (s. sq.)

100 s.sq. = 50 Newton sec. (unit)

⇒ 200 s.sq. = 100 Newton sec.

= Impulse (Ft)

also 100 Newton/sec.

= Impulse (Ft)

also 100 N/sec. = m.v. (mass × velocity)

$$\Rightarrow v = \frac{100}{76} = 1.32 \text{ m./sec.}$$

(B) Impulse = Ft = 62 s.seq = 31 Newton sec.

$$v = \frac{Ft}{m} = \frac{62}{76} = 0.81 \text{ m/sec.}$$

Practicals are possible in Basket ball, Volley ball, High Jump.

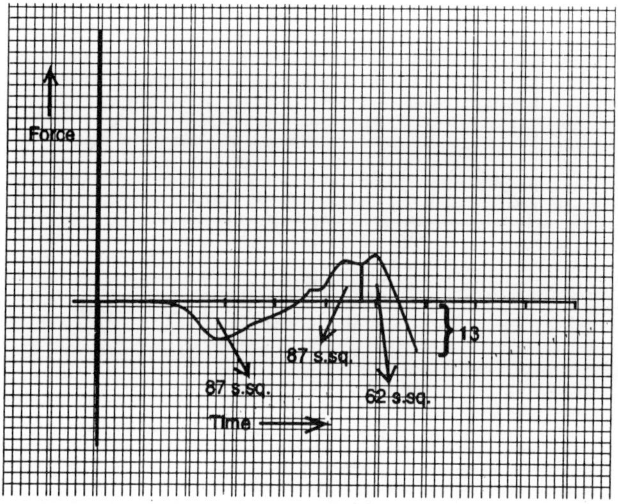

Fig. 13.9:

13.11 EVALUATION OF FORCE TIME CURVE

Acceleration time curve → velocity time curve → distance time

[A, B, and C = Point the athlete

= changes at the point vertical

[point of zeroes

area a_1 = area a_2

area a_4 = area a_3

[for every accleration = decceleration

area a_3 = area a_4]

Maximum increase in the velocity = draw a tangent

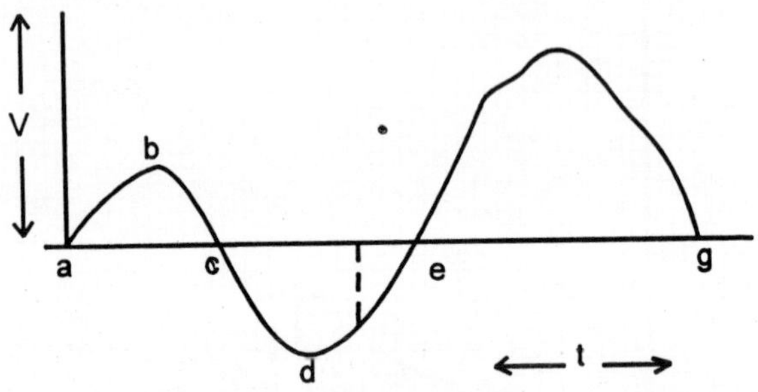

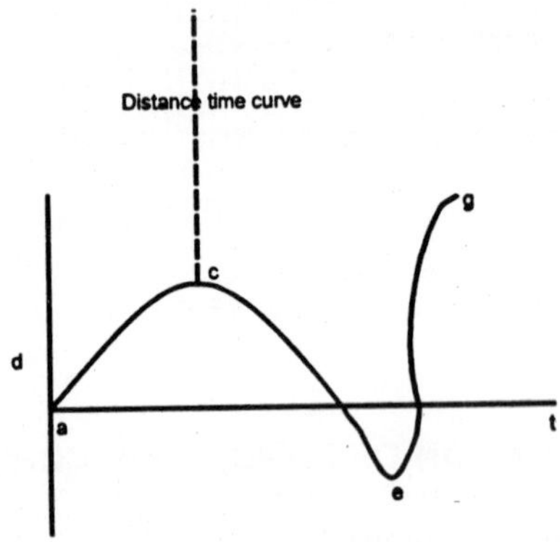

Fig. 13.10:

Distance time curve

How much space has to applied as a partical distance.

Compare the force of people.

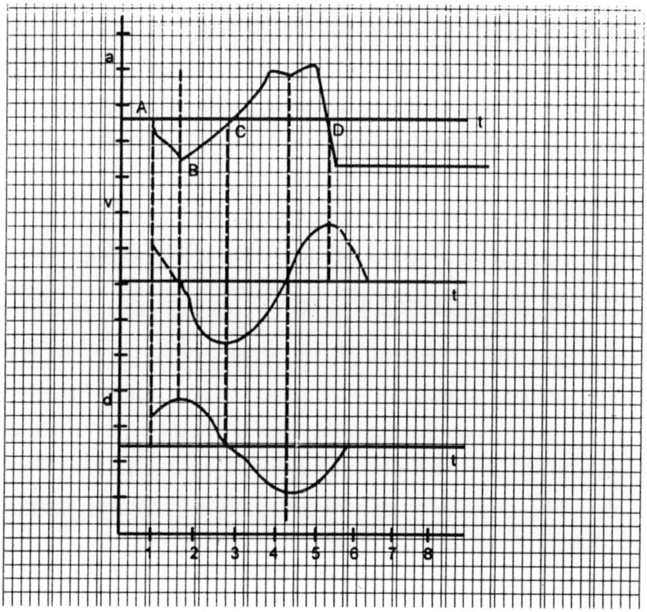

Fig. 13.11:

Boxing F_x, t

Standing High Jump F_y, t

Weight Lifting F_y, t

Foot Ball $F_y \times t$

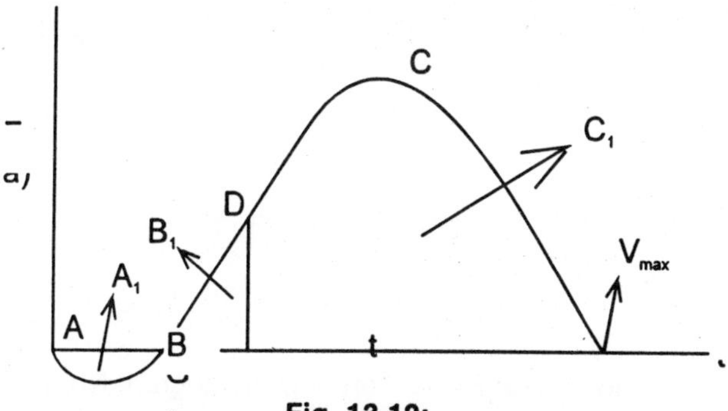

Fig. 13.12:

13.12 EVALUATION OF DYNAMOGRAM NO. 1

The subject was asked to perform a vartical jump on a 'force plateform', which was recorded on a graph paper (Dynamogram). The prcedure followed was:

Registration:

1. Body weight of the subject was ascertained
2. Initial registration was done on the graph.
3. Subject attains a still, standing position on the plateform
4. Vertical jump was performed, while keeping the arms also fixed. (A stick was held on the shoulder) and subjects lands on the plateform again).
5. The performance registered on the dynamogram.

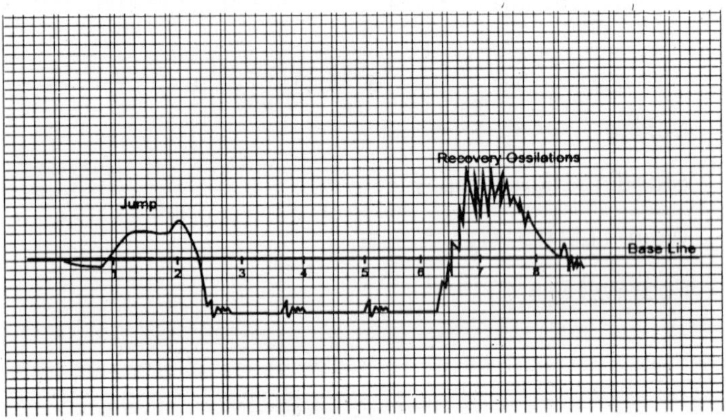

Fig. 13.13: Evaluation of Dynamogram No. 1

Evaluation

1. After getting the dynamogram a base line was attined.
2. The area below and above the base line was ascertained by counting the squares.
3. Area below the base line was deducted from the above one and thus value of jump was recorded in centimeters.
4. The speed of the recorder was assertained.
5. The value of final velocity was attined by the equation.

   $A \times K = m \times v_e$ A = Area m = mass

   k = calibration factor v_e = velocity
6. The value of height for vertical jump was assessed by distance, velocity and time relationship:

$$h = s = \frac{v^2}{2g}$$

Body weight of subject = 86 k.g.

Calculation:–

The area below the base line = 10 m.m

The area above the base line = 446 m.m.

Area for jump = 446 – 10 = 436 m.m.

or $\left(\frac{436}{100}\right)$ = 4.36 square centimeters

Speed machine 50 newton second

For 1 cm. = 50 N sec.

For 4.36 cm. = 4.36 × 50

= 218 Newton second

$\therefore AK = mv_e$ or $mv_e = AK$

86 kg. x v_e = 218

v_e = 2.53 m/sec.

since $h = \dfrac{v^2}{2g}$

$$\therefore = \frac{(2.53)^2}{2 \times 9.81} = \frac{6.4009}{19.62} = .32 \text{ m} = 32 \text{ centimeters}$$

13.13 DYNAMOGRAM NO. 2

Aims

To calculate the value of final velocity (v_f), height of the jump (h), impulse, force f_w and kappa value 'x'.

Procedure

The subject was asked to perform a double leg vertical jump, using both the arms on the force plate form. The method used for registering it was same as explained in dynamometry I.

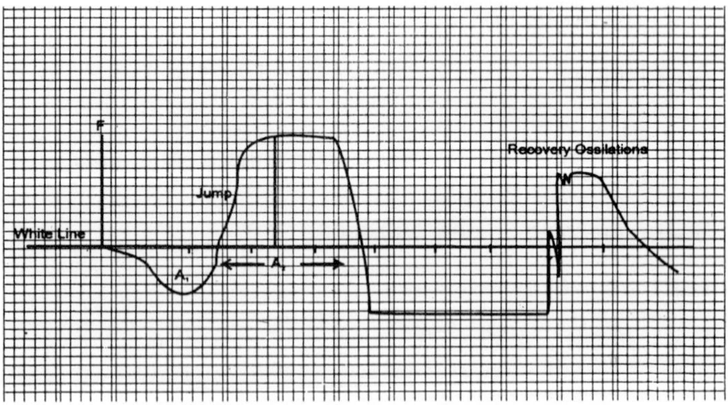

Fig. 13.14: Evaluation of Dynamogram No. 2

Evaluation

Area of Acceleration $= A_2 - A_1$

$= A$

Final velocity $v_f = \dfrac{A \times K.A}{m}$

Distance/ height of jump $= h = \dfrac{v^2}{2g}$

$$\text{Kappa} = \frac{\text{Deceleration}}{\textit{Acceleration}}\left[\frac{A1}{A} = 0.3 \text{ under Normal Conditions}\right]$$

Calculation:-

$A_1 = 206$ sq. m.m Body Weight = 86 kg.

$A_2 = 696$ sq. m.m.

$A = (A_2 - A_1) = 696 - 206 = 490$ sq. m.m. = 4.9 sq. cm.

$v_f = \dfrac{A \times KA}{m}$ $K.A = 1$ cm. = 50 newton sq. cm.

$= \dfrac{4.9 \times 50}{86} = 2.84$ m/sec.

$h = \dfrac{v^2}{2g} = \dfrac{8.090}{19.62} = .41$ metre.

$\text{Kappa} = \dfrac{2.06}{4.09} = 0.5037$

$F_w = 2.3$ cm. x 500 (1 cm^2 = 500 newton on machine)

= 1150 Newton

Figures 13.11, 13.12, 13.13 and 13.14 are provided for evaluation purposes.

Curve Discussion I (A)

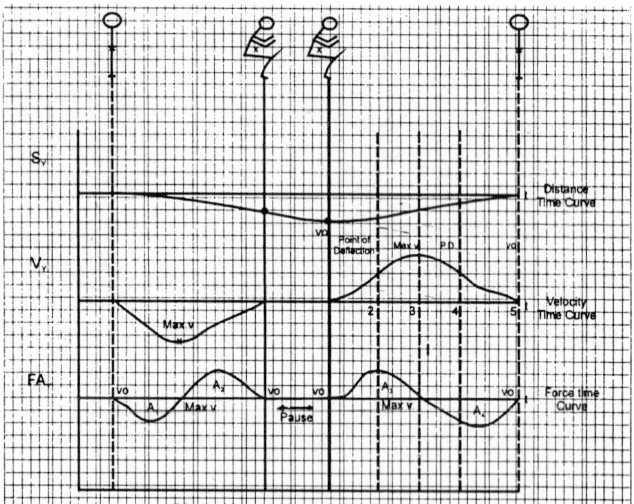

(Vertical jump with a pause in between)

Fig. 13.15:

Curve Discussion I (B)

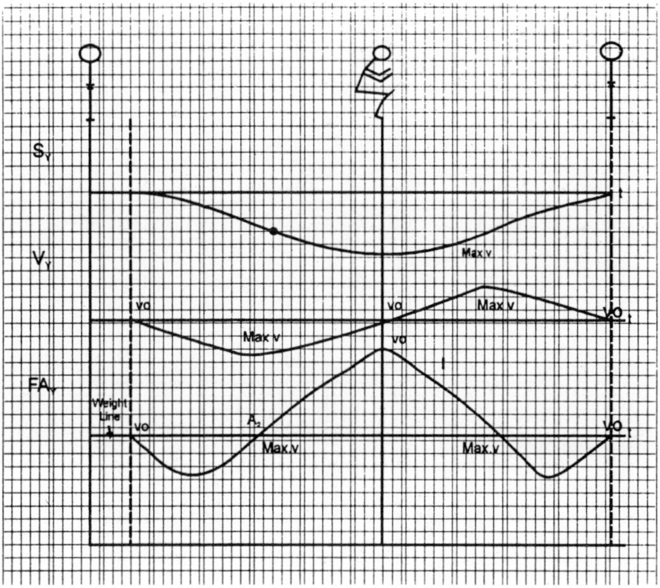

Fig. 13.16:

Curve Discussion: III

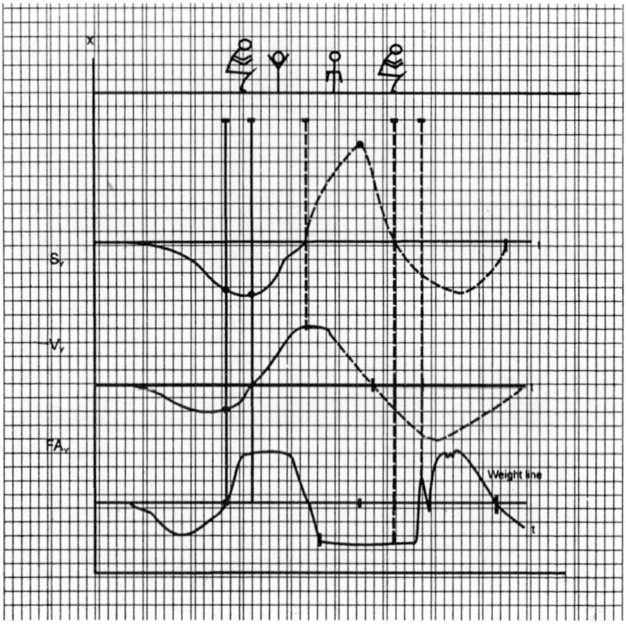

Fig. 13.17:

Curve Discussion: IV

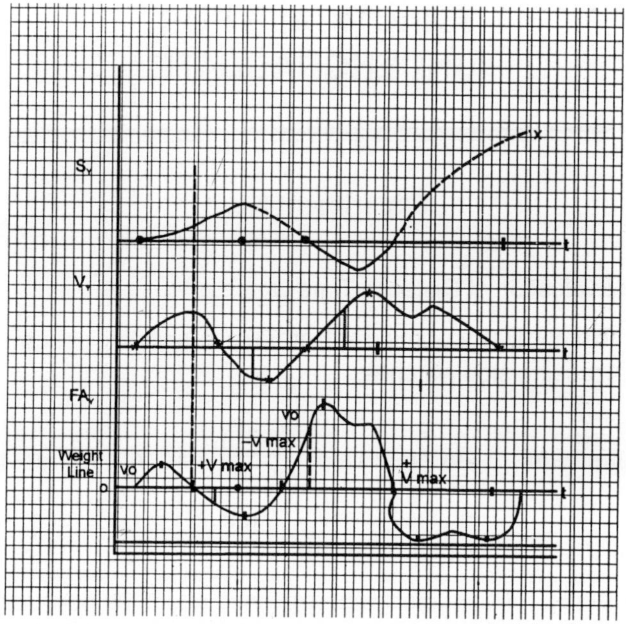

Push Jerk with Downward and Upward Movement

Fig. 13.18:

Chapter-14

PRACTICALS FOR UNDERGRADUATE

Experiment No. 1

Title: Evaluation of Distance Time Graph.

Aim: To draw distance time graph of 100 mts. sprinting performance by using stop watches.

Requirements: Ten stop watch $\left(\frac{1}{100}\right)$ sec, A 100 mts. track (marked at every 10 mts. of intervals) pen, pencil and graph paper.

Procedure: The subject was explained about the performance assistance of 11 co-workers were arranged as per following:—

(a) One starter, (b) At 10 mts. one time keeper with stop watch, (c) At 20 mts. one time keeper with stop watch.

In the same manner one to ten time keeper were allowed to stand at 30 mts., 40 mts, 50 mts, 60 mts, 70 mts, 80 mts, 90 mts, and 100 mts.

All time keeper were placed 2 mts. away from the track of running, standing on the respective marks right angle to the respective 10 mt. lines.

Timing of the subjects were taken from the crossing of either of the toes of each 10 miters cross lines, while performing 100 mts sprinting.

After the competition of 100 mt. sprint by the subject, the split time (arranged) from the stop watch were noted down in the respective chronological manner i.e., 10, 20, 30 .... 100 mts. respectively from the stop watch time and known distance of 10 mt. interval on the graph (distance time graph) was plotted.

Interpretation: Distance time graph have a linearity.

Distance (m)	10	20	30	40	50	60	70	80	90	100
Time (Sec.)	1.85	3.01	4.39	5.21	6.41	7.37	8.54	9.50	10.46	11.72

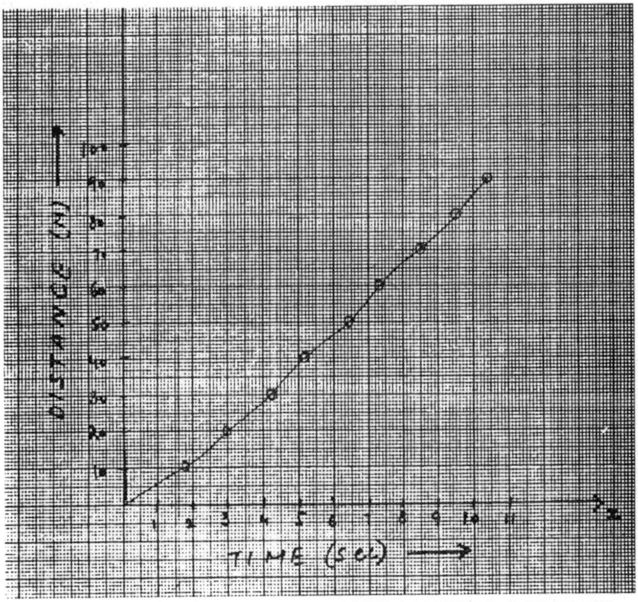

Fig. 14.1: Evaluation of Distance Time Graph

Experiment No. 2

Title: Evaluation of Velocity Time Graph.

Aim: To draw velocity time graph of 100 mts. sprinting performance by using stop watches.

Requirements: Ten stop watches $\left(\frac{1}{100}\right)$ sec., A 100 mt. track marked at every 10 mts. intervals, pen, pencil and graph paper.

Procedure: The subject was explained about the performance assistance of 11 co-workers were arranged as per following:-

(a) One starter (b) At 10 mts. one time keeper with stop watch. (c) At 20 mts. one time keeper with stop watch.

In the same manner, one to ten time keeper were allowed to stand at 30 mts, 40 mts, 50 mts, 60 mts., 70 mts., 80 mts., 90 mts. and 100 mts.

All time keepers were placed 2 mts. away from the running track. Standing on the respective marks, right angle to the respective 10 mts. lane.

Velocity Time Graph (Calculation)

Distance (mts.)	*Average Time (sec.)*	*Split Time in sec.*	*Velocity D/T (m/ sec.)*
10 mts.	1.85 sec. (T_1-T_0)	1.85-0 = 1.85 sec.	10/1.85 = 5.40 m/sec.
20 mts.	3.01 sec. (T_2-T_1)	3.01-1.85 = 1.16 sec.	10/1.16 = 8.62 m/sec.
30 mts.	4.39 sec. (T_3-T_2)	4.39-3.01 = 1.38 sec.	10/1.38 = 7.25 m/sec.
40 mts.	5.21 sec. (T_4-T_3)	5.21-4.39 = 0.82 sec.	10/0.82 = 12.19 m/sec.
50 mts.	6.41 sec. (T_5-T_4)	6.41-5.21 = 1.20 sec.	10/1.20 = 8.33 m/sec.
60 mts.	7.37 sec. (T_6-T_5)	7.37-641 = 0.96 sec.	10/0.96 = 10.42 m/sec.
70 mts.	8.54 sec. (T_7-T_6)	8.54-7.37 = 1.17 sec.	10/1.17 = 8.55 m/sec.
80 mts.	9.50 sec. (T_8-T_7)	9.50-8.54 = 0.96 sec.	10/0.96 = 10.42 m/sec.
90 mts.	10.46 sec. (T_9-T_8)	10.46-9.50 = 0.96 sec.	10.42 = 0.96 m/sec.
100 mts.	11.72 sec. $(T_{10}-T_9)$	11.72-10.46 = 1.26 sec.	10/1.26 = 7.94 m/sec.

Timing of the subject was taken from the crossing of either of the toes while performing 100 mts., sprinting on the respective 10 mt. by the respective time keeper from the start of running to their respective 10 mts.

After the completion of 100 mts. sprint by the subject, the split time (arranged) from the stop watch were noted down from the stop watch time and known distance of 10 mt. interval, the graph (velocity time graph) is plotted by using the formula v = d/t. The velocity time graph was prepared by considering 1cm. = 1 sec. and 1 cm. = 10 mts. On a separate graph paper.

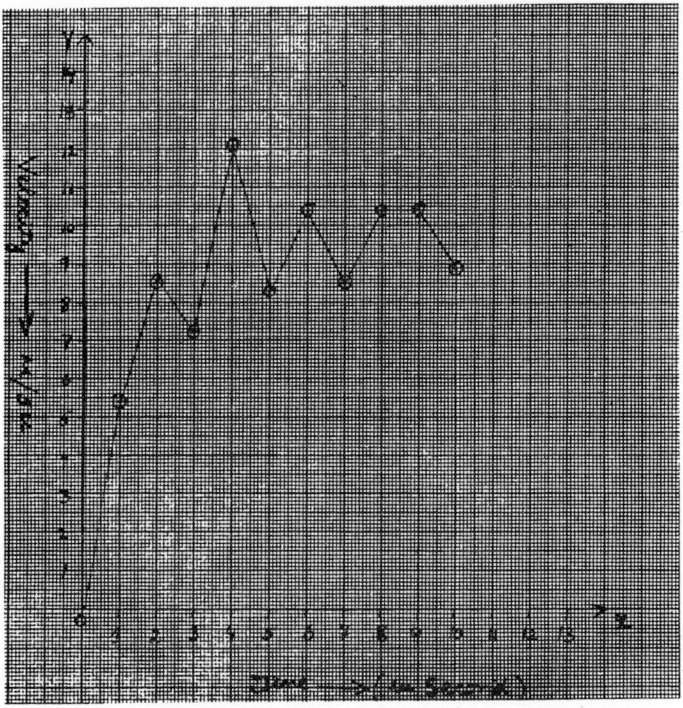

Fig. 14.2: Velocity Time Graph

Interpretation:–

From the graph in the following case, it is well understood that the relationship between time and velocity, that is, at first, velocity gradually increases but on reaching 6.62 sec., it become constant. The acceleration ability is there upto 20 mts. but after that velocity seems to be fluctuating decrease or increase constantly. It means locomotor ability plays important role and the subject obtained at the beginning of 10 mts. but decrease in velocity due to dependence on speed at the end of 100 mts.

Experiment No - 3

Calculation of Acceleration from Velocity Time Graph

Aim:- To draw a acceleration time graph of 100 mts. Sprinting performance with the help of stop watches to evaluate the velocity pattern.

Requirement:- 10 stop watches, 100 mts. lane of a track marked at every 10 mts. of intervals, pen, pencil, calculator, starter and one subject.

Procedure:- The subject was explained and demonstrated about the performance. The assistance of 11 co-workers were taken; one starter, 10 time keepers and one subject.

The time keepers were standing 2 mts. away from the marked line at the right angle to the left side or inside of the track at every 10 mts. interval gap. The timing of each 10 mts. were taken from the starting line when the athlete crosses any of the respective 10 mts. marks. After the completion of 100 mts., sprints, the split time from the stop watches were noted in a chronological manner from 10 mts. to 100 mts. Respectively, from the stop watches, known, distance of 100 mts. with 10 meters interval, velocities were calculated by using the formula →:

$$\text{Velocity} = \frac{\text{distance}}{\text{tine}}$$

Then the time acceleration graph was drawn on a graph paper by considering on scale one cm. is one sec. (1 cm. = 1 sec.) and one cm. is a meter (1 cm. = 1 mts.).

Acceleration Time Graph (Calculation)

Distance (mts.)	*Average Time (sec.)*	*Split Time*	*Velocity (m/sec.)*	*Acceleration (m/sec.²)* $a = \frac{V_f - V_i}{t_f - t_i}$
10 mts.	1.85 sec.	$(T_1 - T_0)$ 1.85 sec.	5.40 m/sec.	$a_1 = \frac{5.40 - 0}{1.85} = 2.9$ m/sec.²
20 mts.	3.01 sec.	$(T_2 - T_1)$ 1.16 sec.	8.62 m/sec.	$a_2 = \frac{8.62 - 5.40}{1.16} = \frac{3.22}{1.16}$ = 2.7 m/sec.²
30 mts.	4.39 sec.	$(T_3 - T_2)$ 1.38 sec.	7.25 m/sec.	$a_3 = \frac{7.25 - 8.62}{1.38} = \frac{1.37}{1.38}$ = -0.99 m/sec.²
40 mts.	5.21 sec.	$(T_4 - T_3)$ 0.82 sec.	12.19 m/sec.	$a_4 = \frac{12.19 - 7.25}{0.82} = \frac{4.94}{0.82}$ = 6.02 m/sec.²
50 mts.	6.41 sec.	$(T_5 - T_4)$ 1.20 sec.	8.33 m/sec.	$a_5 = \frac{8.33 - 12.19}{1.20} = \frac{-3.86}{1.20}$ = –3.2 m/sec.²
60 mts.	7.37 sec.	$(T_6 - T_5)$ 0.96 sec.	10.42 m/sec.	$a_6 = \frac{10.42 - 8.33}{0.93} = \frac{2.09}{0.93}$ = 2.2 m/sec.²
70 mts.	8.54 sec.	$(T_7 - T_6)$ 1.17 sec.	8.55 m/sec.	$a_7 = \frac{8.55 - 10.42}{1.17} = \frac{-1.87}{1.17}$ = –1.598 m/sec.²
80 mts.	9.50 sec.	$(T_8 - T_7)$ 0.96 sec.	10.42 m/sec.	$a_8 = \frac{10.42 - 8.55}{0.96} = \frac{1.87}{0.96}$ =1.95 m/sec.²

(Contd...)

Distance (mts.)	*Average Time (sec.)*	*Split Time*	*Velocity (m/sec.)*	*Acceleration (m/sec.²)* $a = \frac{V_f - V_i}{t_f - t_i}$
90 mts.	10.46 sec.	$(T_9 - T_8)$ 0.96 sec.	10.42 m/sec.	$a_9 = \frac{10.42 - 10.42}{0.96} = \frac{0}{0.96}$ $= 0$ m/sec.^{2}
100 mts.	11.72 sec.	$(T_{10} - T_9)$ 1.26 sec.	7.94 m/sec.	$a_{10} = \frac{7.94 - 10.42}{1.26} = \frac{-2.48}{1.26}$ $= -1.97$ m/sec.^{2}

Then the acceleration were calculated by using the formula of:- $a = \frac{Vf - vi}{t}$

a = acceleration t = time
Vf = Velocity final
vi = Initial velocity

Note:- The unit of acceleration time graph was drawn in mts/100^2. After finding out the acceleration. The acceleration time graph was drawn.

Interpretation:- From the position of the graph, we observe that at the beginning, the acceleration is more. Then after 10 mts., there is decrease in acceleration up to 30 mts. then, the acceleration goes highest start deacceleration. Deacceleration remains till 90 mts. and after that there is a sudden increase in acceleration at the finish.

Velocity Time Graph

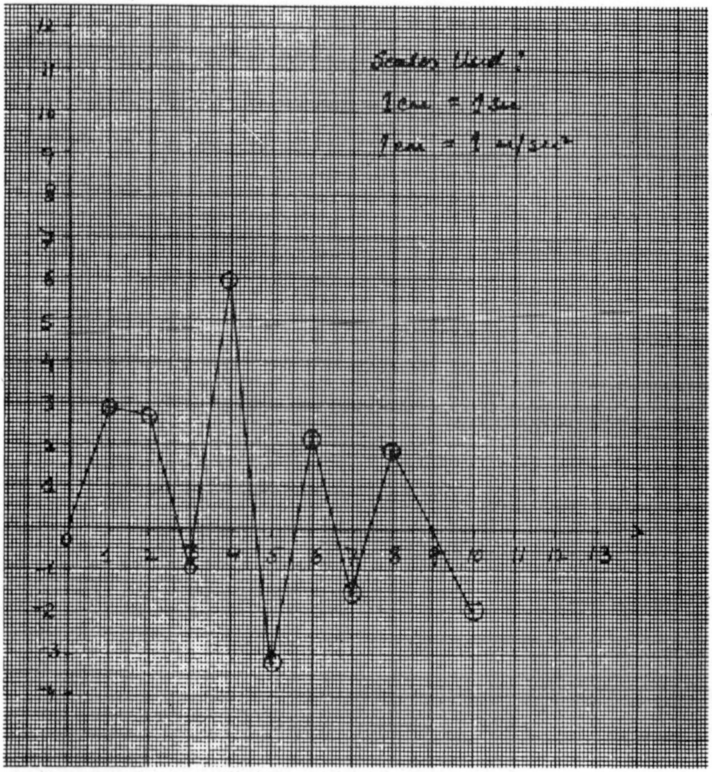

Fig. 14.3: Acceleration from Velocity Time Graph

Experiment No. 4

Aim:- To demonstrate the principle of action and reaction using a turning table.

Equipment:- Turn table, a subject, two dumbells.

Theory:- The experiment is based upon the Newton's IIIrd law, which state that, "To every action there is equal and opposite reaction."

Procedure:- The subject was explained the process and purpose of the experiment. The subject stands on the turn table holding the dumbells in the hand. He was asked to swing his hands towards one direction. His hands moves in the direction of swing while his lower body moves in the oppisite direction.

Observation:- It was observed that when the subject swings his arms towards one direction (action) there was a reaction in the lower body in the opposite direction (reaction).

Conclusion:- It is because of Newton's third law of motion that every action results in equal and opposite reaction. As the arm swing across the body in the one direction an equal and opposite reaction occurs in the rest of the body.

Note: Here the magnitude (equal) of action and reaction could not be demonstrated. Only the direction of action and reaction could be demonstrated.

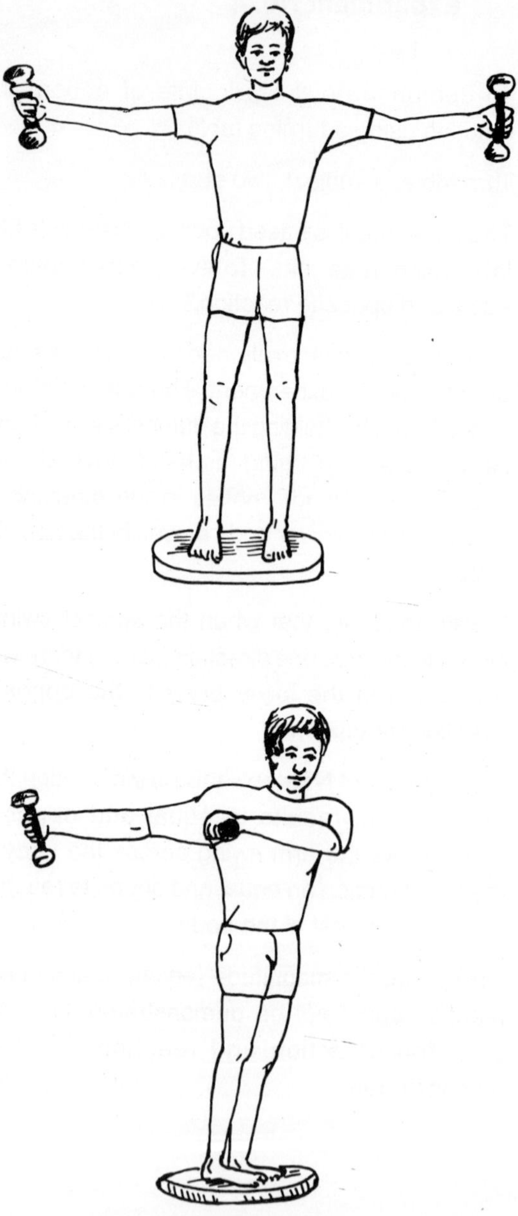

Fig. 14.4: Principle of Action and Reaction

Experiment No. 5

Aim:- To demonstrate the principle of conservation of the angular momentum by using the free turning table.

Material / Requirements:-

Friction less free turning table and the subject.

Theory:- The angular momentum is the product of the angular motions of the mass and it is equivalent of mass and velocity i.e. the movement of inertia.

Inertia is signified by 'I' and an angular velocity by 'W' and angular movement signify with 'IW'.

Procedure:- (1) The subject was explained the aim of the experiment. After that the subject placed himself at the free turning table.

(2) Then it was set in motion while standing on the moving turning table. The subject once stretched out his arms apart side ward and then brought back his arm close to his body by folding them.

(3) The change in velocity in each case was observed closely.

Result:- (1) When the subject was moving with stretched arms, the angular velocity was less.

(2) When the subject was moving with folded arms the angular velocity was mover.

Interpretation:- The velocity of the subject with one stretched arm decreased because his moment of inertia increase as increase in his radius.

Due to the fact, the angular motion or velocity decreased when he folded his arm the angular velocity increases, hence the reduction of moment of inertia occurs, thus an-increase in the angular velocity leads to keep or conserve the angular momentum.

Thus, the moment of inertia about an axis decreases when the mass is concentrated close to the excess of rotation. The moment of inertia of a subject is greater in out stretched arms.

Fig. 14.5: Demonstration of the Principle of Angular Momentum

Experiment No. 6

Title:- Determination of Centre of Gravity of a Badminton Racket. (An Experimental Method).

Equipment:- One Badminton racket, pen, paper etc.

Theory:- The centre of gravity of any object is a point where the centre of mass of that object lies.

Procedure:- (1) The total length of the racket was taken with the help of measuring tape or scale.

(2) The racket was tied up from the centres with the string approximately on the mid of the racket.

(3) The racket was hung freely in the air and the point of balance was located and marked it; that point was the centre of gravity of the racket.

(4) With the help of measuring tape, we measured the length of the racket and also the point of the centre of gravity from both sides.

Result:- (i) The total length of racket was 65 cm.

(ii) Total weight of the racket was 140 grams.

(iii) The distance from top of the racket to the C.G. was 35.5 cm.

Interpretation:- The racket is totally in balanced position when exactly the string was tied at 35.5 cm. from the top and 29.5 cm. from the bottom i.e., the centre of mass of the badminton racket.

In conclusion, we can say that the centre of gravity of the racket is more towards the top then the bottom because of less weight or mass towards wrist, which provide more torque to the face of the racket for stroke.

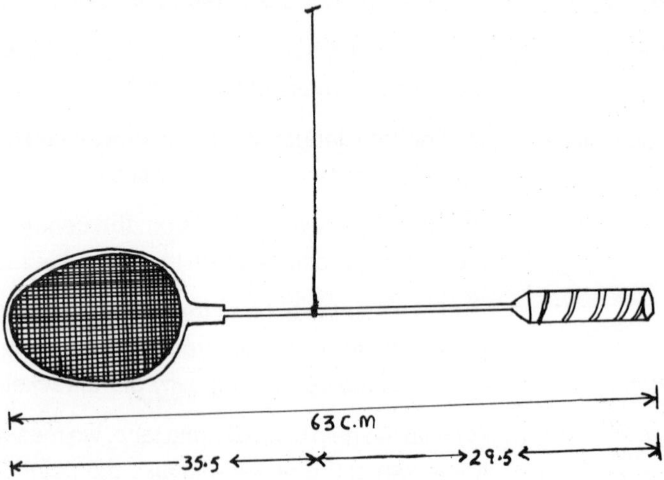

Fig. 14.6: Determination of the C.G. of a Badmintom Racquet

Experiment No. 7

Title: Determination of Centre of Gravity by Reaction Board Method (An Experimental Method).

Aim: Determination of the height of centre of gravity, using a reaction board.

Requirements: (a) Board, of a length 1.83 m.

(b) One subject

(c) Weighing machine

Thoery:

(a) Centre of Gravity:

The centre of gravity is defined as "An imaginary point representing the weight, centre of an object."

or

It is also that point in a body about which all the parts exactly balance each other and can be viewed as the point at which the entire weight of the body may be considered as concentrated.

(b) Reaction Board:

It is the simplest method used to find the centre of gravity of a living human object which involves the use of a so called reaction board.

Procedure: The board is supported horizontally on two "knife edges", one of which rests on a block of wood, the other rests on the platform of a set scales. After the inital reading on the scales (R_1) has been noted, the subject lies in a supine position on top of the board with the soles of the feet pressed against a vertical fitting placed, so that the surface in contact with the feet lies directly above one of the knife edges. The new reading on the scales (R_2) is then noted.

Now, because the board was in equilibrium before the subject lay upon it, and because the board plus-subject is in equilibrium when the subject does lie down, it must be concluded that whatever forces and moments are introduced by the presence of the subject are themselves in equilibrium. The forces introduced are—

(1) The reaction to the increase of weight borne by the platform of the scales and by the wooden block;

(2) The weight of the subject.

Consider the moments of these newly introduced forces about the point A in Fig. 14.6. The moment of the force $R_2 - R_1$ appllied by the platform on the scales to the knife edges. Resting upon it, is 1.83 $(R_1 - R_2)$ N. m or and acts in a clockwise direction. (Note: The horizontal distance between the knife edges of the reaction board is 1.83 m in this case). The moment of the weight force is W × N.m, where x.m is the unknown horizontal distance of the centre of gravity from the point A. This moment acts in a clockwise direction because the body is in equilibrium, rise moments must be equal in magnitude.

That is:

$Wx = 1.8\,(R_2 - R_1)$

Therefore: $x = \dfrac{1.8\,(R_2 - R_1)}{\text{weight of subject}}$

This procedure determines the position of the centre of gravity only in terms of distance from the soles of the feet. Its position laterally (that is, relative to the left or right hand side of the body) or frontally (that is, relative to the front or back of the body) can be obtained using the same basic method, by having the subject asume an errect standing position on the reaction board.

(**Note:** While it would seem reasonable to assume the changing from the position would not alter the location of the centre of gravity relative to the body, it would hardly seem reasonable to expect the same result. When the body position was changed from supine to errect standing mainly because of shifts in the deposition of the body viscera and fluids, the distance of the centre of gravity from the soles of the feet when the subject was standing would probably be slightly less than when he/she adopted a lying position.

Calculation of C.G. by Reaction Board

$$x = \frac{1.8\,(R_2 - R_1)}{W}$$

$$x = \frac{1.8\,(36.300 - 10.700)}{49.50}$$

$$= \frac{1.8\,(25.6)}{49.50}$$

$$= \frac{46.08}{49.50}$$

(Height = 1.61 m)

Where

x = *C.G. of person*

R_2 = *Weight of the subject with the board*

R_1 = *Weight of the board*

Height – 1.61 m.

Result: The height of the centre of gravity of the subject (x) by reaction board method is determined $x = 0.93$ m.

Interpretation: Hence, we conclude, that, while standing, the centre of gravity is above the mid way of total body length to some extend.

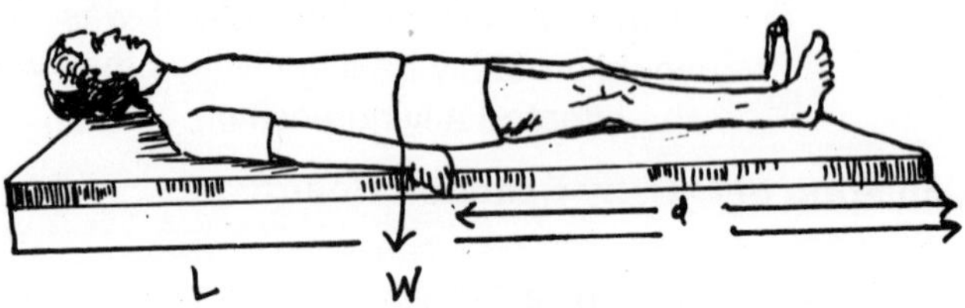

Fig. 14.7: Determination of Centre of Gravity by Reaction Board Method

Experiment No. 8

Title: Determination of Centre of Gravity. By Main Point Method (An Analytical Method).

Aim: Determination of centre of gravity of the given photograph by main point methods with the help of its co-ordinate values on a graph paper.

Requirements: Graph paper, photograph of a player, pen, pencil, scale etc.

Theory: Centre of gravity is a 'point of balance' or "it is an imaginary point of the centre of mass."

Formula:

$$X_{C.G.} = \frac{\sum GT... xnx}{G}, \quad Y_{C.G.} = \frac{\sum GT... yny}{G},$$

Where,

x = Co-ordinated values of the various body parts a. x co-ordinate

G.T. = Relative weight of the body part that is given (in %).

G = Total weight of athlete (in %).

y = Co-ordinated values of the various body parts at *y* co-ordinate.

Calculation of C.G. by Main Point Method

S. No.	Body Parts	Relative Weight	Proximal Distance	Distal Distance	X(cm)	Y (cm)	X.C.G.	Y.C.G.
1.	Head	0.07%	3	4	3.8	15.2	26.6	106.4
2.	Trunk	0.43%	3	2	3.6	12.2	154.8	524.6
3.	Upper Rt. Arm	0.03%	4	5	2.3	15.4	6.9	46.6
4.	Upper Lt. Arm	0.03%	4	5	5.0	15.4	15.0	46.6
5.	Rt. Fore-arm	0.02%	4	5	2.1	17.9	4.2	35.8
6.	Lt. Fore-arm	0.02%	4	5	5.4	17.4	10.8	34.8
7.	Rt. Hand	0.01%	2	3	2.0	19.7	2.0	19.7
8.	Lt. Hand	0.01%	2	3	5.6	19.1	5.6	19.1
9.	Rt. Thigh	0.12%	4	5	2.9	6.9	34.8	82.8
10.	Lt. Thigh	0.12%	4	5	4.0	6.8	48.0	81.6
11.	Rt. Lower Leg	0.05%	4	5	3.0	2.9	15.0	14.5
12.	Lt. Lower Leg	0.05%	4	5	3.7	3.0	18.5	15.0
13.	Rt. Foot	0.02%	4	5	3.3	2.0	6.6	4.0
14.	Lt. Foot	0.02%	4	5	4.1	2.2	8.2	4.4
		1.00 %			50.8	156.1	357.0	1035.1

$$\frac{\text{total } x_{C.G.}}{100\%} = \frac{357.0}{100} = 3.57 \text{ cm}, \ X_{C.G.} = 3.57 \text{ cm.}$$

$$\frac{\text{total } y_{C.G.}}{100\%} = \frac{1035.1}{100} = 10.35 \text{ cm.}, \ Y_{C.G.} = 10.35$$

Working Sheet For Calculating C.G. By Main Point Method

S. No.	*Body Parts*	*Relative Weight*	*Proximal Distance*	*Distal Distance*	*X(cm)*	*Y (cm)*	*X.C.G.*	*Y.C.G.*
1.	Head	0.07%	3	4				
2.	Trunk	0.43%	3	2				
3.	Upper Rt. Arm	0.03%	4	5				
4.	Upper Lt. Arm	0.03%	4	5				
5.	Rt. Fore-arm	0.02%	4	5				
6.	Lt. Fore-arm	0.02%	4	5				
7.	Rt. Hand	0.01%	2	3				
8.	Lt. Hand	0.01%	2	3				
9.	Rt. Thigh	0.12%	4	5				
10.	Lt. Thigh	0.12%	4	5				
11.	Lt. Lower Leg	0.05%	4	5				
12.	Rt. Lower Leg	0.05%	4	5				
13.	Rt. Foot	0.02%	4	5				
14.	Lt. Foot	0.02%	4	5				
		1.00 %						

Methodology:- To determine the point of centre of gravity of the body position of the athlete by this methods the photograph of a player performing the action was pasted on the graph paper.

On the paper, x -axis and y-axis was drawn, the point from the given chart attached were located. From these above located point the perpendicular lines from x-axis and y-axis were drawn. Co-ordinates were measured and written on the test paper. These points were located from the proportional length value from the distal and proximal length. These points were the centre of gravity of every body part.

Result:- The centre of gravity of the subject in x-*axis* is 3.57 cm. and in y-axis is 10.35 cm.

Interpretation:- The centre of gravity of the subject is in middle of the trunk and more towards the upper position of trunk near to the thorax region because subject is in standing position with both the arms stretched upward.

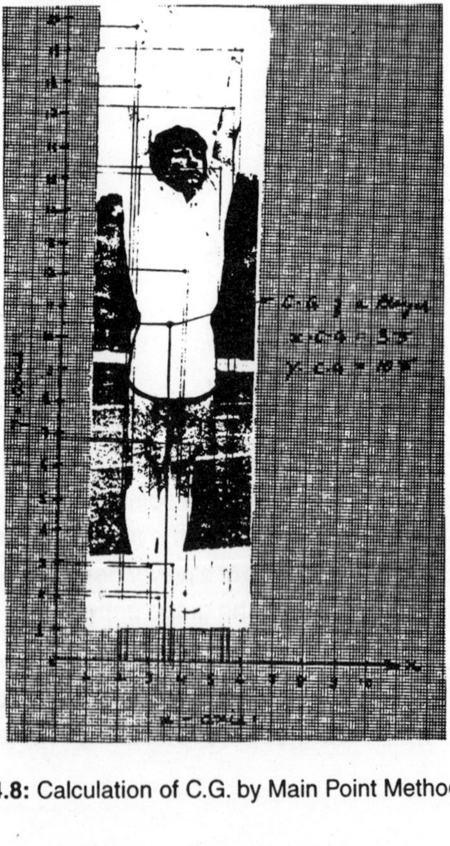

Fig. 14.8: Calculation of C.G. by Main Point Method

Experiment No. 9

Title:- Determination of centre of Gravity by Joint Point Method (An Analytical Method).

Aim:- Determination of a centre of gravity using joint point method of a given photograph.

Material Required:-

Graph paper, photography of a player or an athlete, pen, pencil scale etc.

Theory:- The centre of gravity is the imaginary point representing the weight centre of an object.

Procedure:- (1) Take a photograph of an athlete and paste it on a graph paper.

(2) Draw two axis, y and x on the given graph paper.

(3) Mark the mid point of each of the joints of the body parts of the player.

(4) From that mid-point draw lines perpendicular to the x-axis and y-axis respectively. Then note down the value from the graph table. Then the values of x-axis and y-axis was multiplied by the relative weight of the body parts.

(5) The value of $X_{C.G.}$ and $Y_{C.G.}$, should be added separately and divided by 100 which is the total weight of the body.

Result:- The centre of gravity of the subject in *x-axis* is *3.49* cm. and in *y-axis* is *10.82 cm.*

Intepretation:- The centre of gravity of the subject is in middle of trunk and more towards the upper portion of trunk it is near to the lumber region because subject is in standing position with both the arms stretched upward.

Calculation of C.G. By Joint Point Method

S.No.	Body Part	Relative Weight	X (cm.)	Y (cm.)	X.C.G	Y.C.G
1.	Head	7.0%	4.0	15.1	28.0	105.7
2.	Shoulder Joint (R)	13.6%	2.4	14.6	32.64	198.56
3.	Shoulder Joint (L)	13.6%	4.7	14.5	63.92	197.2
4.	Elbow Joint (R)	2.6%	2.4	16.8	6.24	43.62
5.	Elbow Joint (L)	2.6%	5.0	16.5	13.00	42.9
6.	Wrist (R)	0.8%	2.2	19.1	1.76	15.28
7.	Wrist (L)	0.8%	5.5	18.5	4.4	14.8
8.	Hand (R)	1.0%	2.2	19.1	2.2	19.5
9.	Hand (L)	1.0%	5.5	19.0	5.5	19.0
10.	Hip Joint (R)	16.2%	2.8	9.7	45.36	157.14
11.	Hip Joint (L)	16.2%	4.2	9.7	68.04	157.14
12.	Knee Joint (R)	8.2%	2.6	5.5	21.32	45.1
13.	Knee Joint (L)	8.2%	3.7	5.5	30.34	45.1
14.	Ankle Joint (R)	2.1%	2.7	2.3	5.67	4.83
15.	Ankle Joint (L)	2.1%	3.5	2.4	7.35	5.04
16.	Foot (R)	2.0%	3.0	1.8	6.00	3.6
17.	Foot (L)	2.0%	3.9	1.8	7.8	7.8
		100 %			349.54	1082.31

$$\frac{\text{total } X_{C.G.}}{100} = \frac{349.54}{100} = 3.49,\ X_{C.G.} = 3.49 \text{ cm.}$$

$$\frac{\text{total } Y_{C.G.}}{100} = \frac{1082.31}{100} = 10.82 \text{ cm.},\ Y_{C.G.} = 10.82 \text{ cm.}$$

Working For Determining Centre of Gravity Using Joint Point Method

S.No.	*Body Part*	*Relative Weight*	*X (cm.)*	*Y (cm.)*	*X.G (cm.)*	*Y.G (cm.)*
1.	Head	7.0%				
2.	Shoulder Joint (R)	13.6%				
3.	Shoulder Joint (L)	13.6%				
4.	Elbow Joint (R)	2.6%				
5.	Elbow Joint (L)	2.6%				
6.	Wrist (R)	0.8%				
7.	Wrist (L)	0.8%				
8.	Hand (R)	1.0%				
9.	Hand (L)	1.0%				
10.	Hip Joint (R)	16.2%				
11.	Hip Joint (L)	16.2%				
12.	Knee Joint (R)	8.2%				
13.	Knee Joint (L)	8.2%				
14.	Ankle Joint (R)	2.1%				
15.	Ankle Joint (L)	2.1%				
16.	Foot (R)	2.0%				
17.	Foot (L)	2.0%				
		100 %				

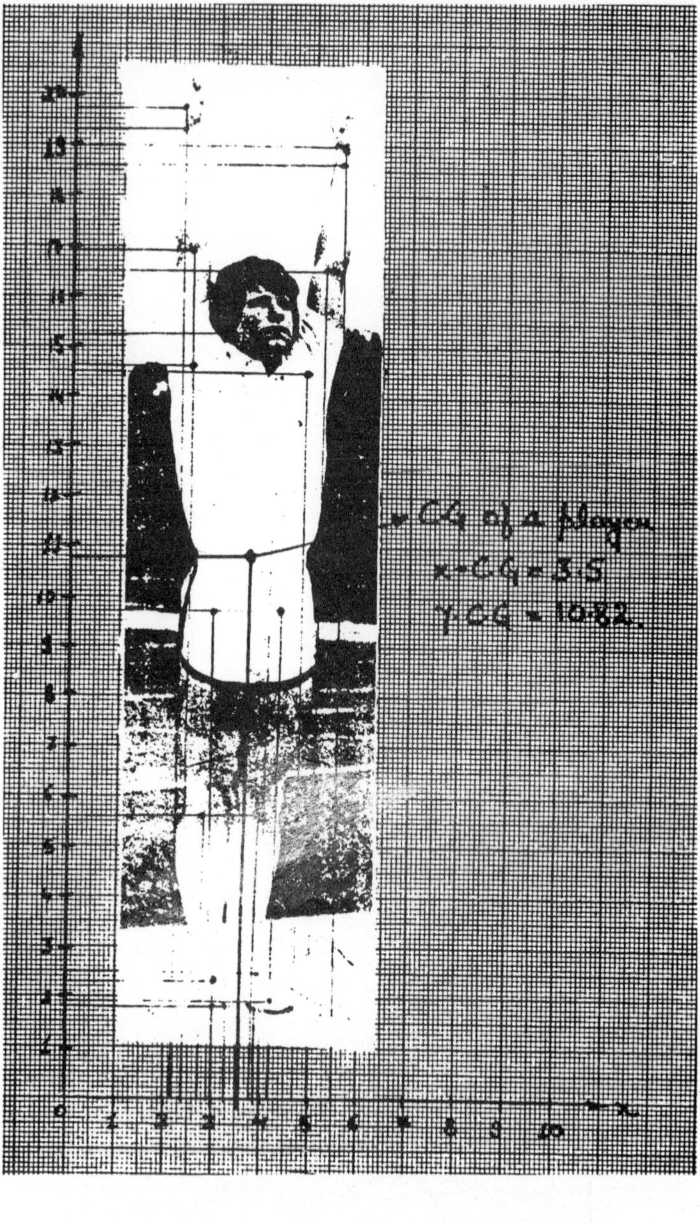

Fig. 14.9: Calculation of C.G. by Joint Point Method

Experiment No. 10

Title:- Determination of Combined Centre of Gravity

Aim:- Determination of Combined Centre of gravity Using joint point method.

Requirement:- Graph paper, photograph of a player, (playing with an object or equipment), pen, pencil, scale etc.

Theory:- Combined centre of gravity, "It is an imaginary point where the weight of an object plus the weight of the body of the player iies.

Formula:-

$$B = \frac{W_2 \times A}{W_1 + W_2} \quad \text{..........................} \quad (i)$$

Where,

B = Displacement of C.G. in x -co-ordinate.

A = Difference between the two bodies in x-values.

W_1 = weight of the subject

W_2 = weight of the object

$$B_1 = \frac{W_2 \times A_1}{W_1 + W_2} \quad \text{..........................} \quad (ii)$$

Where,

B_1 = Displacement of C.G. in y co-ordinate

A_1 = Difference between the two bodies in y-values.

W_1 = Weight of the subject

W_2 = Weight of the object

Methodology:- To determine the combined centre of gravity of the body position of a player, pasted the photograph on a graph paper of a player holding any equipment. Draw x-axis and y-axis respectively to determine the combined C.G. of a playing.

Calculation of C.G. by Joint Point Method

S. No.	Body Parts	Relative Weight	X(cm) Distance	Y (cm)	X.C.G.	Y.C.G.
1.	Head	7%	4.4	12.1	30.8	84.7
2.	Rt. Shoulder Joint	13.6%	3.2	11.5	43.52	156.4
3.	Lt. Shoulder Joint	13.6%	6.0	11.5	81.6	156.4
4.	Rt. Elbow Joint	2.6%	-	8.8	5.2	22.88
5.	Lt. Elbow Joint	2.6%	6.2	9.2	16.12	23.92
6.	Rt. Wrist	0.8%	2.9	10.2	2.32	8.16
7.	Lt. Wrist	0.8%	4.9	10.4	3.92	8.32
8.	Rt. Hand	1.0%	2.9	10.6	2.9	10.6
9.	Lt. Hand	1.0%	4.7	10.7	4.7	10.7
10.	Rt. Hip Joint	16.2%	3.5	8.4	56.7	136.08
11.	Lt. Hip Joint	16.2%	5.2	8.4	84.24	136.08
12.	Rt. Knee Joint	8.2%	3.4	5.3	27.88	43.46
13.	Lt. Knee Joint	8.2%	5.8	5.3	47.56	43.46
14.	Rt. Ankle Joint	2.1%	3.1	2.4	6.51	5.04
15.	Lt. Ankle Joint	2.1 %	6.0	2.3	12.6	4.83
16.	Rt. Foot	2.0%	3.0	1.9	6.0	3.8
17.	Lt. Foot	2.0%	6.2	1.8	12.4	3.6
		100%			444.97	858.43

person marked the mid point of each body parts (joints) and centre of gravity of an implement.

Calculation:- $$B = \frac{W_2 \times A}{W_1 + W_2}$$

A = (x cordinate of subject - x cordinate of object)

$$B = \frac{600 \times (4.44 - 3.6)}{60 + 600}$$

$$B = \frac{600 \times 0.84}{660}$$

$$B = \frac{504}{660} = 0.76 \text{ cm.}$$

$X_{C.G.}$ of a subject + B

4.44 + 0.76 = 5.2 cm.

Displacement of C.G. in x-coordinate is 5.2 cm. from 4.44 cm.

$$B_1 = \frac{W_2 \times A_1}{W_1 + W_2}$$

(A_1 = Y coordinate of subject - Y-coordinate of object)

$$B_1 = \frac{600(8.58 - 11.0)}{60 + 600}$$

$$B_1 = \frac{600(-2.42)}{660}$$

$$B_1 = \frac{-1452}{660}$$

B_1 = -2.2 cm.

$Y_{c.g.}$ of a subject – B1

8.58 – 2.2 = 6.38 cm.

Displacement of C.G. in y-coordinate is 6.38 cm. from 8.58 cm.

Result:- The displacement of centre of gravity of the subject in x-axis is 5.2 cm. and in y-axis is 6.38 cm. from 4.44 cm. and 8.58 cm. respectively.

Interpretation:- The centre of gravity of a subject is lower down and moved towards left direction because the subject is flexed downwards and the object is also in left front of the subject body or chest.

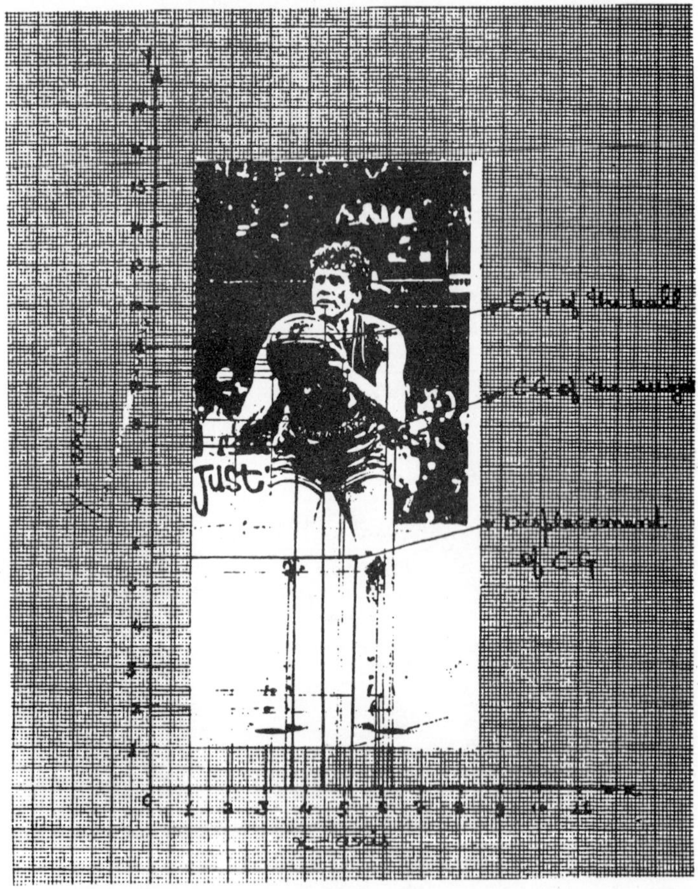

Fig. 14.10: Calculation of C.G. by Combined Joint Method

Experiment No. 11

Aim:- To evaluate dynamogram.

Equipment:- A graph paper and writting meterial and a dynamogram.

Theory:- Dynamograph is a graph that measure the force. The out put of the dynamometer is recorded in graphical form is called dynamogram.

Procedure:- In the graphical form or on the graph a parallel line to the 'X' axis was drawn from the point where the upward trend of the graph starts.

The total number of squares in the area matched by the parallel line and upward curve were counted.

The graph was running at a speed of 10 cm./sec. by the pen recorder, and the given one cm. in the graph was equilient to 500 Newton's standardlization was done as per the following; 1 cm. = 500 × 1 = 500 Newton/sec.

Result:- The total applied force was calculated by putting the formula

$$v = \frac{G \times T}{M}$$

v = velocity

G = 9.81

i.e., force and its output

Total No. of squares = 1900 (was counted by substracting the squares counted in a – b refer Figure 14.11

Mass = 50 kg. (Weight of the person, who performed vertical jump).

Recoder speed = 10 cm./sec.

$\therefore$ 1 cm. = .500 x .1 sec.

= 50.0 Newton/sec.

$$\text{Total no. of square} = \frac{1900}{100} = 19\ \text{cm}^2$$

Total force applied = 50.0 x (19)

= 950 N/sec.

Mometum = m.v

$$v = \frac{\text{force} \times \text{time}}{\text{mass}} = 19$$

G = 9.81 (at Delhi)

$v^2 = 2gh$

$$h = \frac{v^2}{2g} = \frac{(19)^2}{2 \times 9.81}$$

h = 66.4903 centimeters

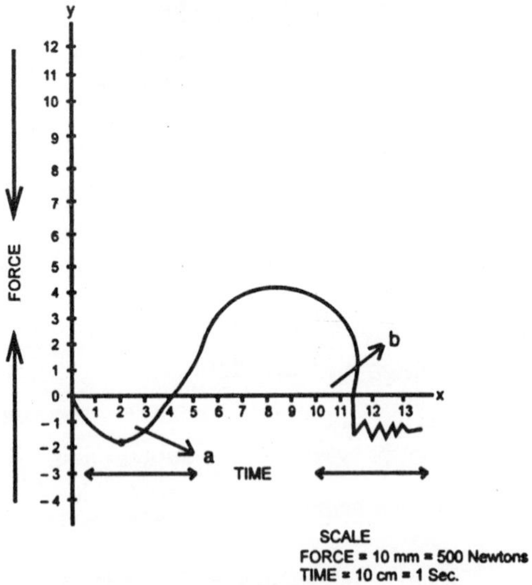

Fig. 14.11: Evaluate Dynamogram

Interpretation:- From the dynamogram it is evident that as force increases – velocity increases and if mass decreases then velocity increases but force decreases, i.e., velocity is directly proportional to the force and inversely proportional to the mass so the calculation reflects that if velocity increases if mass decreases then also greater height is achieved and vice-versa.

Experiment No. 12

Aim:- To evaluate relationship of force, distance, and velocity time graph.

Equipment:- Dynamogram

Theory:- As per graphic presentation with distance and velocity time graph of the paper from the initial point we see that most decline point as the negative sign of force curve. Then we see that point where the force became control i.e., the threshold point of positive and time forces. Now we see that most evident point on the positive sides of forces time curve. We mark at area the positive side of the force time curve equal to the area on the negative side of forces curves.

For velocity time curve from the initial point to the corresponding point of the force where it is neutral is maximum velocity in the negative direction. The point the area on the positive side of force time curve on the negative side the velocity is zero. the ratio between the zero velocity corresponding force time curve point and remaining positive force is 1:2 approx. In distance time curve the velocity is zero or we can say that there is completely sitting position.

The distance is curved maximum in the negative sides its ratio is (1 : 3) approx. where the force is at the peak and in the velocity curve the acceleration at its maximum phase distance also increasing negatively have the ratio is (1 : 3) approx.

Conclusion:- To conclude there are following considerable points that :

(i) The ratio in between negative and positive of force is 1 : 3 respectively.

(ii) The ratio in between negative and positive of distance curve is 1 : 3.

(iii) The raito in between negative and positive of velocity curve is 1 : 3 respectively.

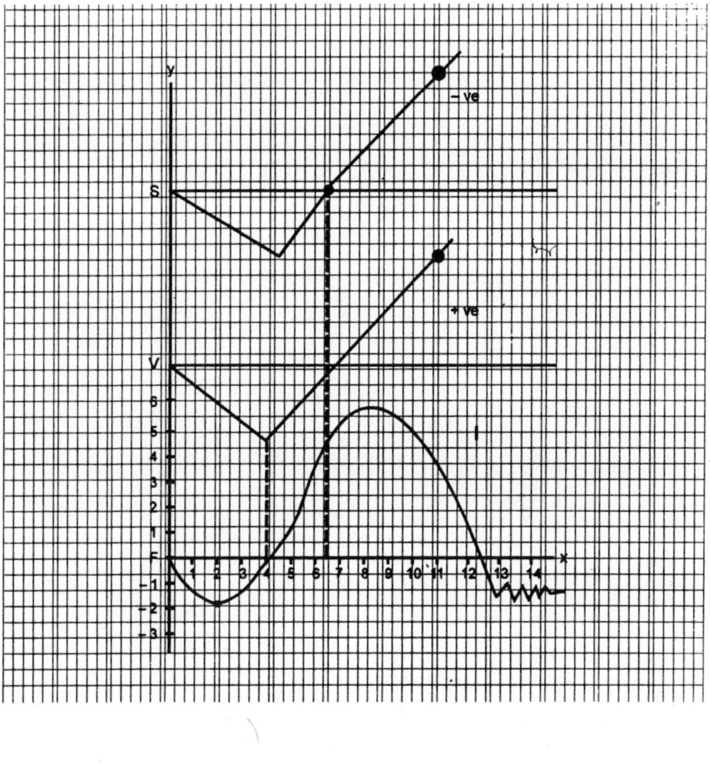

Fig. 14.12:

Chapter-15

SELECTED KINEMATIC AND KINETIC CALCULATION BASED PRACTICAL

ROTATORY MOTION

A discus thrower rotates through 2 and ½ turns before executing. If he takes a time of 1.5 seconds to reach the acceleration point what is his average angular velocity with which he has moved.

Angular distance = 360 x 2½ = 900°

t = 1.5 secs.

$$V = \frac{d}{t} = \frac{900}{1.5} = 600^0\ /\text{sec.}$$

$$\therefore V = \frac{600^\circ}{57.27} = 10.47 \text{ radian/sec.}$$

A discus thrower takes rotation through 900° in 1.2 seconds before he releases his discus. What his *average angular speed*?

$$V = \frac{d}{t} = \frac{900}{1.2} = 750^0 \text{ sec.}$$

$$= \frac{750^\circ}{57.27} = 13.1 \text{ radian/sec.}$$

RELATION BETWEEN ANGULAR AND LINEAR MOTION

(i) A discus thrower rotates through 1 regulation and 140° (500°) before releasing the discus, if he takes a time of one second to cover the total rotation and the distance of the C.G.

of the discus and the medial line of the thrower is 1.10 mts. find out the *linear velocity* of the discus at the time of release.

$$V = \frac{d}{t} = \frac{500^0}{1} = 5000^0 \text{ sec.}$$

or

$$V_r = \frac{500}{57.27} = 8.7305 \text{ radian / sec.}$$

$$V_i = v_r \times r$$
$$= 8.73 \times 1.10$$
$$= 9.603 \text{ m/sec.}$$

(ii) A hammer thrower release the hammer with the velocity of 7 mts/sec. If the distance of the hammer head from the middle axis of the thrower body is 2.20 mts at the time of release find out the average angular velocity of the hammer thrower to come to the release.

$V_l = 7$ m., r = 2.20 m.

$\therefore v_r = ?$

$V_l = V_r \times r$

$$\therefore V_r = \frac{v_l}{r} = \frac{7}{2.20} = 3.18 \text{ raidan/ sec}^2$$

PROJECTILE

(i) A shot putter release the shot at the height of 2.30 m. above the ground with the velocity of 5 m/s. If the shot is released at the angle of 40° with the horizontal — what will be its range and how high the shot will go after the release.

$$R = \frac{v^2 \sin 2\theta}{g}$$

$$h = \frac{v^2 \sin^2 \theta}{2g}$$

$$R = \frac{(5)^2 164 \times 2}{20 \times 32}$$

$$R = \frac{25 \times 1.28}{32}$$

$$R = 1$$

$$= \frac{25 \times (.64)^2}{64}$$

$$= \frac{25 \times .4096}{64}$$

$$= \frac{10.24}{64}$$

L = .16 m.

(ii) A broad jumper takes off at an angle of 26° with the velocity of 7 mts/sec. If it is assumed the jumper is the point object. What will be *his performance* how *high* he will go during his flight.

$$R = \frac{v^2 \sin 2\theta}{g}$$

$$h = \frac{v^2 \sin^2 \theta}{2g}$$

$$R = \frac{(7)^2 \times .43 \times 2}{32} = \frac{49 \times .86}{32} = \frac{42.14}{32} = 1.3 \text{ meters}$$

$$h = \frac{49 \times .43 \times .43}{64} = \frac{49 \times .18}{64} = \frac{9.06}{64} = .14 \text{ ft.}$$

CENTRIFUGAL – CENTRIPITAL – BANKING

(i) A runner is running around a curve with the velocity of 7 m/s, if the runner weights *80 kg* and the radius around which he is running is 20 mts find out the *centrifugal force* developed by the runner while running on the curve and through *what angle* he should lean to avoid the effect of the centrifugal force (cf) force.

$$cf = \frac{wv^2}{gr} \tan\theta$$

$$cf = \frac{80 \times 7^2}{9.81 \times 20} = \frac{3920}{196.2} = 19.97$$

$$\tan\theta = \frac{v^2}{gr} = \frac{7^2}{9.81 \times 20}$$

$$= .2497 = 14^0 \text{ approximately}$$

(ii) A shot putter moves the shot through a distance of 1.6 m in making put and exerts a force of 18 kg. on the shot. How much done *work* before releasing the shot.

W = F × d

= 18 X 1.60

= 28.8 Nm.

(iii) A jumper's weight 70 kg. and *velocity* with which he takes off to cover the maximum horizontal distance 6.54 m/sec. (v) Calculate the KE by the jumper at the time of take off ?

V = 6.5 m/sec. W = 70 k.g.

$$m = \frac{70}{9.81}$$

KE = ½ mv^2

$$= \frac{1}{2} \times \frac{70}{9.81} \times (6.5)^2$$

$$= \frac{35}{9.81} \times 42.25$$

$$= 150.73\,\text{kgm/sec}$$

(iv) A pole voulter clears a height of 4,70 m. If the pole voulter weight 60 kg. What will be *potential energy* developed by him at the pick of his jumps and what will be the *KE* as well as the *velocity* with which he will be hitting the ground at the time of landing.

$$h = 4.70\text{mt. } W = 60 \text{ k.g}, W = mg, m = \frac{W}{g}$$

$$v^2 = 2\,as = 2 \times 9.81 \times 4.70 = x \qquad \left[\begin{array}{l} \because v^2 - u^2 = 2as \\ \text{if } u^2 = 0 \\ \therefore v^2 = 2as \end{array}\right]$$

$$\therefore V = \sqrt{x}\ \text{metres/sec.}$$

$$KE = \frac{1}{2} mv^2 = \frac{60 \times v^2}{2 \times 9.81} = 282\ \text{Ans.}$$

$$PE = mgh = 60 \times 4.70 = 282\ \text{Ans.}$$

BANKING AND CF

(i) An indoor track is to be bent to accommodate a runner with maximum velocity of 10 mts/sec. If the radius (r) around which the curve is to be marked is 30 meters, find out the *angle of the bank* to provide the runner an *automatic lean*. If the runner is running with no lean.

V = 10 m/s, r = 30 mts.

$$\tan\theta = \frac{10^2}{9.81 \times 30} = \frac{100}{294.3} = 3397$$

$$= 18^0 48 \text{ approximately}$$

(Referring log table)

(ii) A cycling velodram is constructed having two radius of 20 mts. and other is 30 mts. It has to accomodate the cyclist upto a velocity of 25 m/sec. *What* should be the *angle* of the banking of the velodram.

When v = 25, r = 20

$$\tan\theta = \frac{25^2}{9.81 \times 20} = \frac{625}{196.2} = 3.18555 = 73°30 \text{ (Approximately)}$$

when v = 25 and r = 30

$$\tan\theta = \frac{25^2}{9.81 \times 30} = \frac{625}{294.3} = 2.1244$$

$\therefore \theta = 64°48'$ (approximately)

1. $c = \frac{P}{W} = \tan\theta =$ coefficient of friction

2. $I = \frac{V_2 - V_1}{U_1 - U_2}$

3. $e = \sqrt{\frac{hb}{hd}}$

FORCE

(i) Sprinter leave the block with the force of 80 k.g. at the time of start, if these forces applied at the angle of 35° to the horizontal, find the *useful component* of the force for the sprinter.

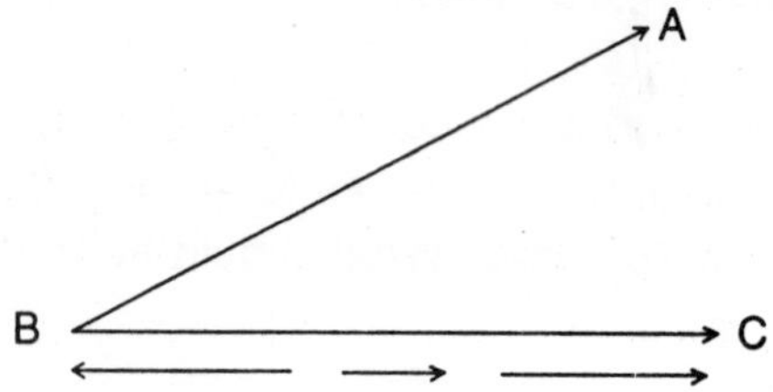

Fig. 15.1:

$$\frac{BC}{AB} = \cos\theta$$

or $BC = AB \cos\theta$

or $BC = 80 \times \cos 35°$

or $BC = 80 \times .8192$

His usefull component = 65.536 kgm

(ii) A high jumper takes off at an angle of 15° with the vertical, if at the time of take off his applied force is 70 kgm on the ground in the opposite direction that of his jump calculate the useful component of this force.

$$\frac{AB}{BC} = \cos\theta$$

$$AB = BL\cos\theta$$

$$AB = 70 \times \cos 15^0$$

$$AB = 70 \times .9659$$

$$= 67.613 \text{ kgm}$$

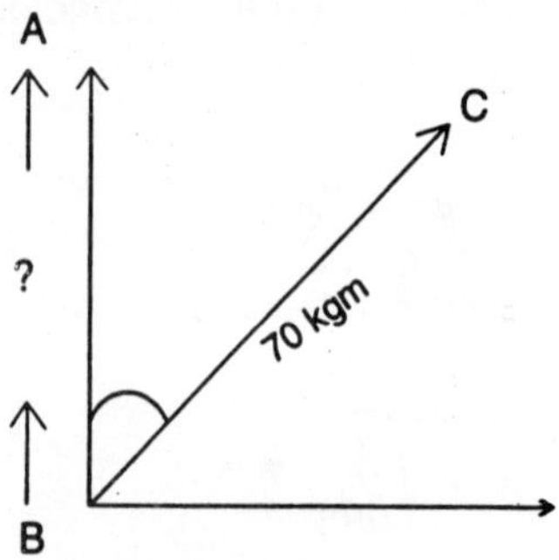

SPEED, VELOCITY, ACCELERATION

(i) A sprinter from the static, acceleration at uniformarly, acceleration rate 2 metres/sec. for a period of 3 seconds what will be his *velocity* at the end of 3rd second and *how much distance* he would have covered till that time?

$s = ?$

$a = 2$ m/s $\quad v = ?$

$t = 3$ sec. Now, $\quad v = u + at$

$v = 0 + 2 \times 3$

$v = 6$ m/s^2

$S = ut + \frac{1}{2} at^2$

or S $0 \times 3 \frac{1}{2} .2(3)^2$

$S = 0 + 9 = 9$ m. distance covered till that times

(ii) A vehicle starts accelarating after accuring a velocity of 5 m/s, at an average rate of acceleration of 7 m/s for 10 sec. What will be the velocity? or of the vehicle at the end of its acceleration and what will be the total distance covered? by the vehicle if it had covered the distance of 30 m prior to accelerating.

$u = 5$ m/s $\quad a = 7$ m/s^2, $t = 10$ s.

then, $v = ?$ and $S = ?$

$V = u + at \qquad S = ut + \frac{1}{2} at^2$

$V = 5 + 7 \times 10 \qquad = 5 \times 10 + \frac{1}{2} at^2$

$= 75$ m/s $\qquad = 5 \times 10 + \frac{1}{2}\, 7 \times (10)^2$

$= 50 + 350$

$= 400$ m.

he has covered already 30 m + 400 m

REFERENCES

Abstracts of the 1st Annual Meeting of American Society of Biomechanics. ***Journal of Biomechanics,*** 1978, 11, 209-215.

Abstracts of the 2nd Annual Meeting of the American Society of Biomechanics. ***Journal of Biomechanics,*** 1979,12, 623-638.

Arian, M. Cinematographic, Electromyographic and Electrogoniometric Techniques for Analysing Human Movement. In J. Wilmore (E.d.), ***Exercise and Sport Sciences Reviews*** (Vol.1). New York: Academic Press.1973.

Allard, P. and Others ed. ***Three Dimensional Analysis of Human Movement,*** *Champaign,* Human Kinetics (1995).

Alexander, R. ***Biomechanics,*** New York: John Wiley & Sons, Inc., 1975.

Alley, L.E. An Analysis of Water Resistance and Propulsion in Swimming the Crawl Stroke. ***Research Quarterly,*** 1952, 23, 253-270.

Alley L.E. Utilization of Mechanics in Physical Education and Athletics. ***Journal of Health, Physical Education and Recreation.*** March1966. pp. 67-70.

Amar J., ***The Human Motor,*** New York E.P. Dutton and Company, 1920.

Asmussen, E. Movement of Man and Study of Man in Motion: A Scànning Review of the Development of Biomechanics. In P.V. Komi (Ed.) ***Biomechanics***

V-A. Baltimore: University Park Press, 1976.

Asumussen.E. and Jorgenson K. (Eds.) ***Biomechanics VI-A, VI-B,*** Baltimore: University Park Press 1978.

Atwater, A. E. Cinematographic Analyses of Human Movement. In J. Wilmore (E.d.), ***Exercise and Sport Sciences Reviews*** (Vol. 1) New York : Academic Press, 1973.

Ayoub, M.A., Ayoub, *M.M.* and Walvekar.A.C.A Biomechanical Model for the Upper Extremity Using Optimization Techniques, ***Human Factors,*** 1974, 16(6), 585 - 594.

Barham, J.N. Organizational Structure of Kinesiology. ***Physical Educator,*** 1963,20(5), 201-21.

Barham, J.N. Toward a Science and Discipline of Human Movement. ***Journal of Health, Physical Education and Recreation,*** October 1966,pp.65-68.

Barham, J.N. ***Mechanical Kinesiology*** St. Louis: C.V. Mosby Company, 1978.

Barham, J.N. and Thomas, W.L. ***Anatomical Kinesiology:*** A Programmed Text, New York: Macmillan Company, 1969.

Barham, J.N.& Wooten, E.P. ***Structural Kinesiology.*** New York Macmillan Company, 1973.

Basmajian. J. V. ***Muscles Alive: Their Functions Revealed by Electromyography.*** Baltimore: Williams and Wilkins Company, 1962.

Basmajian, J.V. The Present Status of Electromyographic Kinesiology. In J. Wartenweiler, E.Joki, and M. Hebbelinck (Eds.) ***Biomechanics* 1.** New York: S., Karger 1968.

Basmajian, J.V. Electromyographic Analysis. In J.M. Cooper (Ed.), ***Selected Topic On Biomechanics,*** Chicago : Athletic Institute, 1971.

Basmajian, J.V. ***Muscles Alive: Their Function Revealed by Electromyography*** (4th Ed.). Baltimore: Williams and Wilkins Company, 1978.

Bergemann, B.W. Three Dimensional Cinematography: A Flexible Approach. ***Research Quarterly,*** 1947, 45, 302-309.

Bleustein, J.L. (E.d), ***Mechanics and Sport,*** New York: American Society of Mechanical Engineers, AMD, Vol. 4, 1973.

Boccardi, S. Biomechanics and Medical Rehabilitation: Introductory Remarks In S. Cerquiglini, A., Venerando, & J. Wartenweiler (Eds). ***Biomechanics-III.*** Baltimore: University Park Press, 1973.

Boccardi, S., Frigo, C., Rodano, R, Santambrogio. C., and Pedotti, A. Analysis of Some Vector Diagrams in J. Terauds & G.G.Dales (Eds.) ***Science in Athletics,*** Del Mar, Chafn Academic Publishers, 1979.

Bootzin, D. & Muffley, H.C.(Eds.) ***Biomechanics,*** New York: Plenum Press, 1969.

Bowen, W.P. ***The Action of Muscles in Bodily Movement and Posture.*** Springfield Mass: F.A. Bassette Co., 1912.

Bowen, W.P., and Mckenzie, K.L. ***Applied Anatomy and Kinesiology — The Mechanism of Human Movement,*** Philadelphia: Lea & Febizer, 1917.

Boysen, J.P., Francis, P.R. and Thomas, R.A Interactive Computer Graphics in the Study of Human Body Planar Motion under Free Fall Conditions. ***Journal of Biomechanics*** 1977, 10, 783-787.

Braun, G.L. Kinesiology From Aristotle to the Twentieth Century ***Research Quarterly,*** 1941, 12, 163-173.

Broer, M.R. ***Efficiency of Human Movement*** Philadelphia: W.B Saunders Company, 1960.

Broer, M.R. ***An Introduction to Kinesiology*** Englewood Cliffs. N.J.: Prentice Hall Inc., 1968.

Broer, M.R. Laboratory Experiences: ***Exploring Efficiency of Human Movement.*** Philadelphia : W.B. Saunders Company, 1973.

Broer, M.R. and Houtz, S.J. ***Patterns of Muscular Activity in Selected Sport Skills: An Electromyogrphic Study.*** Springfield, Ill: Charles C Thomas, 1967.

Broer, M. R. and Zernicke, R.F. ***Efficiency of Human Movement*** (4th ed.), Philadelphia : W.B.Saunders Company, 1979.

Brunnstrom, S. ***Clinical Kinesiology*** Philadelphia: F.A. Davis Company, 1962.

Brunnstrom, S. ***Clinical Kinesiology*** (3rd ed.). Revised by R. Dickinson. Philadelphia: F.A. Davis Company, 1972.

Bunn J.W. ***Scientific Principles of Coaching.*** Englewood Cliffs, N.J.: Prentice-Hall Inc.,1955.

Bunn J.W. ***Scientific Principles of Coaching*** (2nd ed.) Englewood Cliffs. N.J.: Prentice-Hall. Inc.,1972.

Burke R.K. The future of Kinesiology at the Undergraduate Level In C.J. Diffman and R.G. Sears (Eds.), ***Proceedings, Kinesiology: A National Conference on Teaching.*** Urbana- Champaign: University of Illinois,1977.

Burstein. A.H. and Wright T.M. ***Fundamentals of Orthopaedic-Biomechanics.*** Baltimore: Williams and Wilkins Company, 1994.

Butler, D.L. Grood, E.S, Noyes, F.R. and Zernicke, R.F. Biomechanics of Ligaments and Tendons. In R.S. Hutton (Ed.), ***Exercise and Sport Science Reviews***

(Vol. 6) Philadelphia: Franklin Institute Press, 1979.

Byars, E.F., Contini, R. and Roberts, V.L. ***Biomechanics Monograph*** New York: American Society of Mechanical Engineers, 1967.

Cavanga, G.A. Force Platforms as Ergometers. ***Journal of Applied Physiology,*** 1975,39, 174-179.

Cavangh, P.R. ***Recent Advances in Instrumentation and Methodology of Biomechanical Studies.*** In P. V Komi (Ed.), Biomechanics V. B. Baltimore: University Park Press. 1976.

Cavangh, P.R. Instrumentation and Methodology of Applied and Pure Research in Biomechanics. In F. Landry and W.A.R. Orban (Eds.), ***Biomechanics of Sports and Anthropometry.*** Miami:Symposia Specialists, 1978. (a).

Cavangh, P.R.A Technique for Averaging Centre of Pressure Paths from a Force Platform. ***Journal of Biomechanics,*** 1978, 11,487-491, (b).

Cerqinglini, S., Venerando. A. and Wartenweiler, J. (Eds.), ***Biomechanics III.*** Baltimore: University Park Press, 1973.

Contini, R. Preface. ***Human Factors,*** 1963, 5(5), 423-425.

Contini, R., and Drills, R. ***Biomechanics: Applied Mechanics Reviews,*** 1954,7(2), 49-52.

Contini, R. and Drills, R. Biomechanics. In H.N. Abramson, H. Libowitz, J.M. Crowley, and S. Juhasz (Eds.), ***Applied Mechanics Surveys.*** Washington, D.C.: Spartan Books 1966(a).

Contini, R. and Drillis, R. Kinematic and Kinetic Techniques in Biomechanics. In F. Alt(Ed.) *Advances in* ***Bioengineering and Instrumentation*** 1. New York:

Plenum Press, 1966 (b).

Cooper, J.M. (Ed.), ***Selected Topics on Biomechanics,*** Chicago : Athletic Institute, 1971.

Cooper, J.M. Research in Biomechanics In L.L. Gedvilas (Ed.), ***Proceedings of the 79th Annual Meeting of NCPEAM,*** Hot Speing, Ark, 1976.

Cooper, J.M. The Historical Development of Kinesiology with Emphasis on Concepts and People In C.J. Dillman and R. G. Sears (Eds.), ***Proceedings, Kinesiology: Annual Conference on Teaching.*** Urbana-Champaign: University of Illinois, 1977.

Cooper, J.M. and Glassow, R.B. ***Kinesiology.*** St. Louis C.V. Mosby Co., 1963.

Cooper, J.M. and Glassow, R.B. ***Kinesiology*** 4th ed. St. Louis: C.V. Mosby Co., 1976

Crpeluwe, B.V. Ed. and Atha, J. (ed.) ***Current Research in Sports Biomechanics,*** Paris Kargler, 1987.

Cureton, T.K. Mechanics and Kinesiology of Swimming (the Crawl Flutter Kick). ***Research Quarterly,*** 1930, 1, 87-121.

Cureton, T.K. Physics Applied to Physical Education. ***Journal of Health and Physical Education,*** January 1932, pp. 23-25.

Cureton, T.K. Mechanics of the Shot Put. ***Scholastic Coach,*** March 1935. pp. 7-10 (a).

Cureton, T.K. Mechanics of the High Jump. ***Scholastic Coach,*** April 1935, pp. 9-12 (b).

Cureton, T.K. Elementary Principles and Techniques of Cinematographic Analysis as Aids in Athletic Research. ***Research Quarterly.*** 1939,10, 3-24.

Cureton T.K. Review of a Decade of Research in Aquatics at Springfield College, 1929-1939. ***Research Quarterly,*** 1940, 11, 68-79.

Daniels, L. and Worthingham, C. ***Muscle Testing Techniques of Manual Examination.*** Louson, W.B. Erunders Com.

Dempster, W.T. Space Requirement of the Seated Operate ***WADC Technical Report*** 55-159. Wright-Patterson Air Force Base, Ohio, 1955.

Dempster, W.T. Free-body Diagrams as an Approach to the Mechanics of Human Posture and Motion in F.G. Evans (Ed.). ***Biomechanical Study of the Musculoskeletal System.*** Springfield. Ill: Charles C. Thomas Publisher, 1961.

Dillman, C.J. Muscular Torque Patterns of the Leg During the Recovery Phase of Sprint Running (Doctoral Disseration. Pennsylvania State University 1970). ***Dissertation Abstract International,*** 1971, 32,222 A. (University Microfilms No. 71-16.952).

Dillman, C.J. Kinematic Analysis of Running. I.J. Wilmore & J. Keogh (Eds.), ***Exercise and Sport Science Reviews*** (Vol.3). New York: Academic Press, 1975.

Dillman, C.J., and Nelson, R.C. The Mechanical Energy Transformations of Pole Vaulting with a Fibreglass Pole. ***Journal of Biomechanics,*** 1068, 1, 175-183.

Dillman, C.J., and Sears R.G.(Eds.), Proceedings, ***Kinesiology*** A National Conference on Teaching Urbana-Champaign: University of Illinois, 1977.

Dorland's ***Illustrated Medical Dictionary*** (24th ed.). Philadelphia: W.B. Saunders Company, 1965.

Drillis, R.Folk Norms and Biomechanics ***Human Factor*** 1963, 5 (5), 427-442.

Duvall, E.N. Kinesiology: ***The Anatomy of Motion.*** Englewood Cliffs. N.J.: Prentice-Hall. Inc., 1959.

Dyson, G.H.G. ***The Mechanics of Athletics,*** London: University of London Press, 1962.

Dyson, G.H.G. ***The Mechanics of Athletics*** (7th ed.),New York: Holmes and Meier Publishers, 1977.

Elftman, H. ***The Measurement of the External Force in Walking Science,*** 1938,88, 152-153.

Elftman, H. Forces and Energy Changes in the Leg During Walking: ***American Journal of Physiology,*** 1939, 125, 339-356 (a).

Elftman, H. The Function of Muscles in Locomotion ***American Journal of Physiology,*** 1939,125,357-366 (b).

Eneka, R.M. ***Neuromechanical Basis of Kinesiology,*** Champaign, Human Kinetics Book, 1988.

Eneka, R.M. ***Neuromechanical Basis of Kinesiology*** IInd Ed. Champaign: Human Kinetics, 1994.

Falkel, J.E. and Others ***Shoulder Injuries*** Baltimore: Williams and Wilkines, 1988.

Fenn, W.O. Mechanical Energy Expenditure in Sprint Running as Measured by Moving Pictures. ***American Journal of Physiology,*** 1929,90,343-344.

Fenn, W.O. Frictional and Kinetic Factors in the Work of Sprint Running. ***American Journal of Physiology,*** 1930,92, 583-611 (a).

Fenn, W.O. Work Against Gravity and Work Due to Velocity Changes in Running. ***American Journal of Physiology,*** 1930,93,433-462 (b).

Fenn, W.O.A Cinematographic Study of Sprinters. ***Scientific Monthly,*** 1931,32,346-354.

Frankel, V.H., and Burstein, A.H. ***Orthopaedic Biomechanics.*** The Application of Engineering to the Musculo-Skeletal System, Philadelphia: Lea & Febiger, 1970.

Frankel, V.H. and Hong.Y. Recent Advances in the Biomechanics of Sport Injuries. ***Acta Orthopaedia Scandinavica,*** 1975, 46 (3), 484-497.

Fung, Y.C.(Ed.), ***Biomechanics,*** New York: American Society of Mechanical Engineers, 1966.

Fung,Y.C. Biomechanics-its Scope, History, and Some Problems of Continuum Mechanics in Physiology. ***Applied Mechanics Reviews,*** 1968,21(1), 1-20.

Fung, Y.C., Perrone, N., and Anliker, M.(Eds.), ***Biomechanics —Its Foundations and Objectives*** Englewood, Cliffs. N.J.: Prentice-Hall, Inc., 1972.

Furusawa, K., Hill. A.V., and Parkinson, J.L. The Dynamics of Sprint Running. ***Proceedings of the Royal Society of London,*** 1927,102B, 29-42 (a).

Furusawa, K. Hill. A.V., and Parkinson, J.L. The Energy Used in Sprint Running. ***Proceedings of the Royal Society of London,*** 1927, 102B, 43-50 (b).

Garrett, G.E. Development of Biomechanics. ***Proceedings of the NAPECWINCPEAM National Conference,*** Denver, Colo., 1978.

Glassow, R.B. ***Fundamentals of Physical Education.*** Philadelphia: Lea & Febiger, 1932.

Glassow, R.B. ***Laboratory Manual for Functional Kinesiology*** Madison, Wisg Kramer business service, 1950.

Glassow, R.B. Modern Dance and Kinesiology, **Journal of Health,Physical Education, Recreation,** January 1966, pp.65-68.

Glassow, R.B., Hubbard, A.W., and Brown F., Photographical and Cinematographical Research Methods. In M.G. Scott & T.K. Cureton (Eds.), ***Research Methods Applied to Health, Physical Education, and Recreation.*** Washington D.C.: American Association for Health, Physical Education and Recreation, 1949.

Grabiner, M.D. Ed. ***Current Issue in Biomechanics*** Champaign, Human Kinetics Publishers, 1993.

Gray, J. ***Animal Locomotion*** London: Weidenfield and Nicolson, 1968.

Gregor, R.J., and Zernicke, R.F. Kinesiology as a life science: An Undergraduate Curriculum. In C.J. Dillman & R. G. Sears (Eds.), ***Proceedings Kinesiology: A National Conference on Teaching*** Urbana Champaign: University of Illinois. 1977.

Grieve, D. W. A Device Called the Pologon for the Measurement of the Orientation of Parts of the Body Relative to a Fixed External Axis. ***Journal of Physiology,*** 1969, 201-70p.

Grieve, D.W. The Dynamics of Lifting. In R.S. Hutton (Ed.), ***Exercise and Sport Sciences Reviews*** (Vol.5), Santa Barbara:L Journal Publishing Affiliates, 1978.

Grieve, D. W. Miller d.i. Mitchelson, D. Paul., J.P. & Smith, A.J. ***Techniques for the Analysis of Human Movement*** Princeton: Princeton Book Company Publishers, 1975.

Groves, R. and Camasone, D.N. ***Concepts in Kinesiology*** Philadelphia: W.B Saunders Company, 1975.

Harries, R. W. ***Kinesiology Workbook and Laboratory Manual*** Boston: Houghton Mifflin Company, 1977.

Hart, I. ***The Mechanical Investigations of Lenardo da Vinci*** (2nd Ed.),Borkeley: University of California Press, 1963.

Hatze, H. Letter: The Meaning of the Term "Biomechanics." ***Journal of Biomechanics,*** 1974,7,189-190.

Hatze, H. Biomechamical Aspects of a Successful Motion Optimization. In P.V. komi (Ed.), ***Biomechanics*** V.B. Baltimore: University Park Press, 1976 (a).

Hatze, H. The Complete Optimization of a Human Motion. ***Mathematical Bioscience,*** 1976, 28. 99-135 (b).

Hawley, G. ***The Kinesiology of Corrective Exercise.*** Philadelphia: Lea & Febiger, 1937.

Hawley, G. ***An Anatomical Analysis of Sports,*** New York: A.S. Barnes and Company, Inc., 1940.

Hay, J.G. An investigation of Mechanical Efficiency in Two Styles of High Jumping (Doctoral Dissertation. University of Iowa, 1967). ***Dissertation Abstract International,*** 1968, 28A, 3004A. (University Microfilm no. 68-937).

Hay, J.G. ***Biomechanics of Sports Technique*** *3rd Ed.,* New Jersey: Prentice Hall, 1985.

Hay, J.G. Experiments in the Mechanics of Physical Activities. ***Journal of Health, Physical Education and Recreation,*** January 1969, pp. 89-90.

Hay, J.G. ***The Biomechanics of Sports Techniques.*** Englewood Cliffs. N.J.: Prentice-hall, Inc., 1973 (a).

Hay, J.G. The Centre of Gravity of the Human Body. In C.J. Widule (Ed.), ***Kinesiology III.*** Washington.D.C.: American Association for Health, Physical Education and Recreation, 1973 (b).

Hay, J.G.(Ed.), ***Kinesiology IV,*** Washington, D.C.: American Association for Health, Physical Education and

Recreation, 1974 (a).

Hay, J.G. Moment of Inertia of the Human Body. In J.G. Hay (Ed.), ***Kinesiology IV.*** Washington, D.C.: American Association for Health, Physical Education and Recreation, 1974 (b).

Hay, J.G. Biomechanical Aspects of Jumping. In J. Wilmore & J. Keogh (Eds.), ***Exercise and Sport Science Reviews*** (Vol.3), New York Academic Press,1975.

Hay, J.G. ***The Biomechanics of Sports Techniques*** (2nd ed.), Englewood Cliffs, N.J.: Prentice-Hall, Inc.,1978 (a).

Hay, J.G. Biomechanics: The Present and Future State of the Discipline in the United States and Other Western Countries. In F. Landray & W.A.R. Orban (Eds.), ***Biomechanics of Sports and Kinanthropopometry,*** Miami; Symposia Specialists, Inc., 1978 (b).

Hay, J.G, Wilson,B.D.,& Depena, J.Identification of the Limiting Factors in the Performance of a Basic Human Movement. In P.V. Komi (Ed.), ***Biomechanics*** V.B. Baltimore: University Park Press,1976.

Hay, J.G. ***Biomechanics of Sports Technique*** 3rd ed., New Jersey, Prentice-Hall, 1985.

Herman, R.M., Grillner, S., Stein, P.S.G. and Stuart, D.G. (Eds.), ***Neural Control of Locomotion.*** New York: Plenum Press, 1976.

Hildebrand, M. Symmertrical Gaits of Primates. ***American Journal of Physical Anthropology,*** 1967, 26,119-130.

Higgins, J.R. ***Human Movement:*** An Integrated Approach, St. Louis: C.V. Mosby Company, 1977.

Hill, A.V. Scientific Study of Athletes ***Scientific American,*** 1926, 134, 224-225.

Hill, A.V. Are Athletes Machines? ***Scientific American.*** 1927, 137, 124-126 (a).

Hill, A.V. ***Muscular Movement in Man,*** New York: McGraw Hill Book Co., 1927 (b).

Hinson, M.M. ***Kinesiology,*** Dubuque, Iowa: Wm.C.Brown Company Publishers, 1977.

Hirt, S. What is Kinesiology? A Historical Review, ***Physical Therapy Review,*** 1955,35(8),419-426.

Hochmuth, G. and Others ***Biomechanics of Athletic Movement.*** Berlin: Sporturlag, 1984.

Hopper, B. J. ***The Mechanics of Human Movement,*** New York: American Elsevieer Publishing Company. Inc.,1973.

Hubbard, A.W. An Experimental Analysis of Running and of Certain Fundamental Differences Between Trained and UnTrained Runners. ***Research Quarterly,*** 1939, 10, 28-38 (a).

Hubbard, A.W. Muscular Force in Reciprocal Movements. ***Journal of General Psychology.*** 1939, 20, 315-325 (b).

Hubbard, A.W. Photography. In M.G. Scott (Ed.), ***Research Methods in Health, Physical Education and Recreation*** (2nd ed.), Washington, D.C.: American Association for Health, Physical Education and Recreation, 1959.

Hubbard, A.W. Homokinetics: Muscular Function in Human Movement. In W.R. Johnson (Ed.), ***Science and Medicine of Exercise and Sports.*** New York: Harper & Brothers Publishers, 1960.

Hubbard, A.W., Cureton, T.K., Huelster, L. & Keller, L.F. Research Methods in the Mechanics of Sports and Physical Education Activities. In M.G. Scott & T.K. Cureton(Eds.), ***Research Method Applied to***

Health, Physical Education and Recreation, Washington, D.C.: American Association for Health, Physical Education and Recreation, 1949.

Hubbard, A.W., Cureton, T.K., Steinhaus, A.H. & Stetson, R.H. Method of Research in Experimental Kinesiology. In M.G.Scott & T.K. Cureton (Eds.). ***Research Methods Applied to Health, Physical Education and Recreation.*** Washington, D.C.: American Association for Health, Physical Education and Recreation, 1949.

Hudgins, C.V. The Incidence of Muscular Contraction in Reciprocal Movement under Condition of Changing Loads. ***Journal of General Psychology,*** 1939, 20,327-338.

Inman, V.T., Saunders, J.B., Dec. M., & Abbott, L.C. Observation on the Function of the Shoulder Joint. ***Journal of Bone and Joint Surgery,*** 1944,26,1-30.

James, S.L. Biomechanical and Neuromuscular Aspect of Running. In J. Wilmore (Ed.), ***Exercise and Sport Sciences Reviews*** (Vol.1), New York: Academic Press, 1973.

Jensen, C.R., & Schultz, G.W. ***Applied Kinesiology: The Scientific Study of Human Performance*** (2nd ed.), New York: McGraw-Hill book Co.,1977.

Jensen, R.K Structure, Function and Sport- A Biomechanical Viewpoint. ***Canadian Journal of Applied Sports Science,*** 1978, 3, 5-8.

Kanwar, R.C. ***Shareer Rachna Kriya Vijyan Avema Swasthya Shiksha.*** Nagpur: Amit Brothers Publications.

Karpovich, P.V. and Karpovich, G.P. Electrogoniometer: A New Device for Study of Joints in Action, ***Federation Proceedings,*** 1959, 18,79.

Kasvand, T., Milner, M., and Rapley, L.F.A Computer-based System for the Analysis of Some Aspects of Human Locomotion. ***In Human Locomotion Engineering.*** London: Institution of Mechanical Engineers,1974.

Kelly, D.L. ***Kinesiology: Fundamentals of Motion Description.*** Englewood Cliffs, N.J.: Prentice-Hall. Inc., 1971.

Kenedi, R.M. and Cowden, J.M. Ed. ***Bedsore Biomechanics.*** London, University Park Press, 1976.

Komi, P.V. (Ed.), **Biomechanics V.A./V.B.** Baltimore: University Park Press, 1976.

Kranz, L.G. ***Kinesiology Manual.*** St. Louis: C.V.Mosby Company, 1948.

Krause, J.V. & Barham, J.N. ***The Mechanical Foundations of Human Motion: A Programmed Text*** St. Louis: C.V. Mosby Company, 1975.

Kreighbaum, E. and Barthels, K.M. ***Biomechanics. A Qualitative Approach for Studying Human Movement*** 2nd ed. New York: Macmillan Publishing Co., 1985.

Laban, R.'Reprint'. ***Mastery of Movement*** 4th ed. V.K. Northcate House, 1988.

Landry, F. and Orban,W.A.R.(Eds.), ***Biomechanics of Sports and Kinanthropometry.*** Miami: Sysposia Specialists, Inc., 1978.

Leveau, B., Williams and Lissner. ***Biomechanics of Human Motion*** (2nd ed.), Philadelphia: W.B. Saunders Company, 1977.

Lewillie. L., and Clarys, J.P. (Eds.). ***Biomechanics of Swimming.*** Brussles: University Libre de. Bruxellesx, 1971.

Lindholm, L.E. An Opto-Electonic Instrument for Remote On-line Movement Monitoring. In R.C. Nelson and C.A.

Morehouse (eds.), ***Biomechanics IV.*** Baltimore : University Park Press,1974.

Liptovetz, F.J. ***Basic Kinesiology*** Minneapolis: Burgess Publishing Co.,1952.

Lissner, H.R. Biomechanics—What is it? In E.F. Byars, R. Contini, & V.L. Robers (eds.), ***Biomechanics Monograph,*** New York: American Society of Machanical Engineer, 1967.

Locke, L.F. Kinesiology and the Profession, ***Journal of Health, Physical Education and Recreation,*** September 1963, pp. 69-71.

Locke, L.F. Kinesiology: A Word or a Discipline? ***Journal of Health, Physical Education and Recreation,*** June 1967, p.50.

Logan, G.A. and Mckinney, W.C. ***Anatomic Kinesiology*** *(2nd ed.).* Dubuque, Iowa: W.M.C. Brown Company Publishers, 1977.

Luttgens, K., & Zernicke.R.F. Structure and Organization of Kinesiology Biomechanics Content at Undergraduate Level. In C.H. Dillman & R.C.Sears (Eds.), ***Proceedings, Kinesiology A National Conference Teaching.*** Urbana-Champaign: University of Illinosis,1977.

Luttgens, K. and Others ***Kinesiology Scientific Basis of Human Motion,*** 8th ed. Dubuque, W.C.B. Publisher, 1992.

Mac Conaill, M.A., & Basmajian, J.V. ***Muscles and Movement: A Basis for Human Kinesiology.*** Baltimore: Williams & Wilkins Company, 1969.

Mac Conaill, M.A.,& Basmajian, J.V. ***Muscles and Movement: A Basis for Human Kinesiology.*** *(2nd ed.).*

Huntington, New York: Robert E.Krieger Publishing Company, 1977.

Mann, R.V., and Sorensen, H.C. Investigating Complex Movement by Computer Simulation. Application to the Flip Long Jump. ***Research Quarterly,*** 1979,50,36-50.

Margaria, R. ***Biomechanics and Energetic of Muscular Exercise.*** Oxford Charendon Press, 1979.

Matake, T. On the New Force Plate Study. In P.V. Komi (Ed.), ***Biomechanics*** *V.B.* Baltimore: University Park Press, 1976.

McCloy, C.H. The Organization and Teaching of Apparatus Work and Tumbling. ***Journal of Physical Education,*** March-April 1937, pp.60-62.

McCloy, C.H. The Mechanical Analysis of Motor Skills. In W.R.Johnson (Ed.), ***Science and Medicine of Exercise and Sports.*** New York: Harper and Brothers Publishers,1960.

McLaughlin, T.M., Dillman, C.J., and Lardner, T.J. Biomechanical Analysis with Cubic Spline Functions. ***Research Quarterly,*** 1977,48,569-582.

Miller, D.I. AComputer Simulation Model of the Airbone Phase of Diving. (Doctoral Dissertation. Pennsylvania State University, 1970). ***Dissertation Abstracts International,*** 1971, 32.223-A, (University Microfilm No. 71-16.644).

Miller, D.I. A Computer Simulation Model of the Airbone Phase of Diving in J.M.Cooper (Ed.). ***Selected Topics on Biomechanics.*** Chicago: Athletic Institute,1971.

Miller, D.I. Biomechanics of Swimming. In I. Wilmore & J. Keogh (Eds.), ***Exercise and Sport Sciences Reviews*** (Vol. 3.) New York: Academic Press, 1975 (a).

Miller, D.I. Computer Simulation of Human Motion. ***In Techniques for the Analysis of Human Movement.*** Princeton: Princeton Book Company Publishers,1975 (b).

Miller, D.I. Biomechanics of running — What should the Future Hold. ***Canadian Journal of Applied Sport Science,*** 1978, 3, 229-236.

Miller, D.I. Modelling in Biomechanics: An Overview. ***Medicine and Science in Sports,*** 1979,11(2),115-122.

Miller, D.I., and Nelson. R.C. ***Biomechanics of Sport: A Research Approach.*** Philadelphia: Lea & Febiger, 1973.

Miller, D.I. and Petak, K.L. Three — Dimensional Cinematography. In C.J.Widule (Ed.). ***Kinesiology III.*** Washington, D.C.: American Association for Health, Physical Education and Recreation, 1973.

Mitchelson, D.L. An Opto — Electronic Technique for Analysis of Angular Movement. In S.Cerquiqlini, A. Venerando. & J. Wartenweiler (Eds.), ***Biomechanics III.*** Baltimore: University Park Press, 1973.

Mitchelson, D.L. Recording of Movement without Photography, ***Techniques for the Analysis of Human Movement.*** Princeton: Princeton Book Company Publishers, 1975.

Miyashita, M. Biomechanics of Sports from the View Point of Methodological Advances, In P.V.Komi (Ed.), ***Biomechanics V.B.,*** Baltimore: University Park Press, 1976.

Moon, S.D. and Santer, S.L. Ed. ***Beyound Biomechanics*** Psychological Aspects of Musculo-Skeletal Disorders in Office Work: London, Tayler & Francis, 1996.

Morehouse, L.E., & Cooper, J.M. ***Kinesiology*** St. Louis: C.V. Mosby Company, 1950.

Muybridge, E. ***The Human Figure in Motion.*** New York: Dover Publications, 1955.

Muybridge, E. ***Animal in Motion.*** New York: Dover Publications, 1957.

Nelson, R.C.(Ed.), ***Biomechanics Conference: Proceedings,*** University Park, Pa: Biomechanics Laboratory, 1971 (a).

Nelson, R.C. ***Biomechanics of Sport: An Overview.*** In J.M. Cooper(Ed.), Selected Topics on Biomechanics, Chicago: Athletic Institute, 1971 (b).

Nelson, R.C. *The New World of Biomechanics of Sport.* ***Proceedings of the 75th annual meeting of Ncpeam,*** New Orieans, 1972.

Nelson, R.C. Morehouse C.A. (Eds.). ***Biomechanics IV.*** Baltimore: University Park Press, 1974.

Northrip. J.W. Logan. G. A. and Mckinney, W.C. ***Introduction to Biomechanics Analysis of Sport.*** Dubuque, Iowa: Wm. C. Brown Company Publishers, 1974.

Northrip, J.W. Logan. G.A. & Mckinney, W.C. ***Introduction to Biomechanics Analysis of Sport*** (2nd ed.). Dubuque, Iowa: Wm. C. Brown Company Publishers, 1979.

O'Connell, A.L. (Ed.), ***Kinesiology Review 1968.*** Washington D.C.: American Association for Health, Physical Education and Recreation, 1968.

O'Connell, A.L., and Cardner, E.B. ***Understanding the Scientific Bases of Human Movement.*** Baltimore: Williams & Wilkins Company, 1979.

Oseid, S. and Carlson, K.H. ***Children and Exercise XIII*** Champaign, Human Kinetics Books (International Series on Sport Sciences V.19), 1989.

Paul, J.P. Instrument for Force Measurement. ***Techniques for the Analysis of Human Movement.*** Princeton: Princeton Book Company, Publishers, 1975.

Payne, A.H., Slater, W.J. and Telford, T. The Use of a Force Platform in the Study of Athletic Activities: A Preliminary Investigation, ***Ergonomics,*** 1968, 11(2), 123-143.

Pezzack, J. C. Norman, R.W. & Winter, D.A. An Assessment of Derivation Determining Techniques Used for Motion Analysis. ***Journal of Biomechanics,*** 1977,10,377-382.

Plagenhoef, S. C. An Analysis of the Kinematics and Kinetics of Selected Symmetrical Body Actions (Doctoral Dissertation, University of Michigan, 1962). ***Dissertation Abstract International,*** 1963, 23, 3227. (University Microfilms No. 63-430).

Plagenheof, S.C. Computer Programs for Obtaining Kinetic Data on Human Movement. ***Journal of Biomechanics,*** 1968, 1, 221-234.

Plagenheof. S.C ***Patterns of Human Motions: A Cinematograph Analysis.*** Englewood Cliffs. N.J.: Prentice- Hall, Inc.,1971.

Poortman, J.R.Ed. ***Principles of Exercise Biochemistry*** Paris: Karger, 1988.

Posse, N. ***The Special Kinesiology of Educational Gymnastics*** Boston: Lothrop, Lee, and Shepard Company, 1890.

Remay, M.R. A Simulation of the Running Long Jump. In J.L. Bleustein (Ed.). ***Mechanics and Sports,*** New York:

American Society of Mechanical Engineers, 1973.

Ramey, M.R. Force Plate Designs and Applications. In J. Wilmore & J. Keogh, (Eds.), ***Exercise and Sport Science Reviews*** (Vol.3). New York: Academic Press, 1975.

Rasch, P.J. Notes Toward a History of Kinesiology. Parts I,II and III. ***Journal of the American Osteopathic Association.*** 1958, 58,572-574: 641-644: 713-714.

Rasch. P.J. & Burke, R.K. ***Kinesiology and Applied Anatomy: The Science of Human Movement.*** Philadelphia: Lea & Febiger, 1959.

Rasch, P.J. & Burke, R.K. ***Kinesiology and Applied Anatomy: The Science of Human Movement*** *(4th ed.).* Philadelphia: Lea & Febiger, 1971.

Rasch, P.J. and Bruke, R.K. ***Kinesiology and Applied Anatomy: The Science of Human Movement*** *(6th ed.).* Philadelphia: Lea & Febiger, 1978.

Reed, D.J. and Reynolds, P.J. A Joint Analog Detector. ***Journal of Applied Physiology,*** 1969, 27, 745-748.

Reid, J.G.Muscles in Action: Use of Videotape and Electromyography. ***Journal of Health, Physical Education, Recreation,*** September 1971, pp. 61-62.

Roberton, M.A. Development Kinesiology. ***Journal of Health, Physical, Education and Recreation,*** October 1972, 65-66.

Roberts, V.L. Editorial ***Journal of Biomechanics:*** 1979,12,573.

Sarget, R.N. and Brunk. J.W. Relation of Perceptual-motor Ability and Intellectual Ability in Elementary School Children. ***Perceptual and Motor Skills,*** 1967, 24, 967-970.

Schmide, R. A. Schema Theory of Describe Motor Skill Learning. ***Psychological Review,*** 1973, 82, 225-260.

Schmide.R. A . Schema Theory of Discrete Motor Skill Learning. ***Psychological Review,*** 1975,82,225-260.

Scott, M.G. ***Analysis of Human Motion. A Textbook in Kinesiology,*** New York: Appleton-Century-Crofts. Inc., 1942.

Scott, M.G. (Ed.). ***Research Methods Applied to Health, Physical Education and Recreation*** *(2nd ed.).* Washington.D.C.: American Association for Health, Physical Education and Recreation, 1959.

Scott, M.G. and Cureton,T.K. (Eds.) ***Research Methods Applied to Health, Physical Education and Recreation,*** Washington. D.C.: American Association for Health, Physical Education and Recreation, 1949.

Seireg, A. Leonardo do Vinci –The Bio-Mechanician. In D. Bootzin & H. C. Muffley (Eds.), ***Biomechanics,*** New York: Plenum Press, 1969.

Seashote, H.G. Some Relationships of Fine and Gross Motor Abilities. ***Research Quarterly,*** 1942, 15, 259-274.

Scheme, A. Herh Paugle. R. The Effects of a Structured Physical Education Program on Physician Interrelationship and Self Concept Development of Educable Retarded Boys, IMRID Gearge Peabody College, 1966.

Shapiro, R. Direct Linear Transformation Method for Three–Dimensional Cinematography. ***Research Quarterly,*** 1978, 49, 197-205.

Shaw, Dhananjoy ***Pedagogic Kinesiology.*** Delhi: Sports Publication, 1998.

Shaw, Dhananjoy ***Fundamental Statistics in Physical & Sports Sciences.*** Delhi: Sports Publication, 1998.

Shaw, Dhananjoy ***Biomechanics and Kinesiology of Human Motion.*** Delhi: Khel Sahitya Kendra, 1998.

Shaw, Dhananjoy and Shalini "Evolution of Biomechanics As A Subject of 21st Century" Ed. Shukla, N.B. ***Recent Trend in Physical Education and Sports,*** Varanasi: Indian Society of Sports Scientist, 1999.

Shaw, Dhananjoy and Tomar, Rakesh ***Doctoral Research in Physical Education and Its Sciences in Developed Countries.*** Delhi: Khel Sahitya Kendra, 2000.

Shaw, Dhananjoy and Gambhir, Shalini ***Encyclopaedia of Sports Injuries And Indian Sports Persons.*** Delhi: Khel Sahitya Kendra, 2000.

Sigerseth, P.O. and Mcloy, C.H. Electromyographic Study of Selected Muscles Involved in Movements of Upper Arm at Scapulohumeral Joint. ***Research Quarterly,*** 1956, 27, 409-417.

Skarstrom, W. ***Gymnastic Kinesiology.*** Springfield Mass.: F.A. Bassette Co., 1909.

Slater-Hammel, A.T. An Action Current Study of Contraction–Movement Relationships in Golf Stroke. ***Research Quarterly,*** 1948, 19, 164–177.

Smith, A.J. Photographic Analysis of Movement. ***Techniques for the Analysis of Human Movement.*** Princeton: Princeton Book Company, Publishers,1975.

Spence, D.W. ***Essentials of Kinesiology: A Laboratory*** Philadelphia: Lea & Febiger, 1975.

Sperry, R.W. Action Current Study in Movement Coordination, ***Journal of General Psychology,*** 1939,20, 295-313.

Steindler, A. ***Mechanics of Normal and Pathological Locomotion In Man.*** Springfield, Ill: Charles C. Thomas Publisher, 1935.

Steindler, A. What has Biokinetics to offer to the Physical Educator: ***Journal of Health and Physical Education,*** November 1942, 507-509, 555-556.

Steindler, A. ***Kinesiology of the Human Body Under Normal and Pathological Conditions.*** Springheld, Ill: Charles C. Thomas Publisher, 1955.

Stochmever, S. A. An Interpretation of the Approach of Root to the Treatment of Neuro-Muscular Dysfunction. ***American Journal of Physical Medicine,*** 1967, 46, 900–956.

Suger, R.N. Interrelationship of Physical Perceptual–Motor and Academic Achievement Variables in Elementary School Children. ***Perceptual and Motor Skills,*** 1968, 27, 1323-1332.

Suger, R.N. Interrelationship of Physical Perceptual — Motor and Academic Achievement Variables in Elementory School Children. ***Perceptual and Motor Skill,*** B1968, 27, 1323-1332.

Terauds, J. (Ed.) ***Science in Racquet Sports*** Del Mar. Calif.: Academic Publishers, 1979 (a).

Terauds, J. (Ed.) ***Science in Skiing, Skating, and Hockey*** Del Mar. Calif.: Academic Publishers, 1979 (b).

Terauds, J. (Ed.) ***Science in Sports: Open Papers*** Del Mar. Calif.: Academic Publishers, 1979 (c).

Terauds, J. (Ed.) ***Science in Weight Lifting.*** Del Mar. Calif.: Academic Publishers, 1979 (d).

Terauds, J. and Bedingfield, E.W. (Eds.), ***Swimming III,*** Baltimore: University Park Press, 1979.

Terauds, J. and Dales, G.G. (Eds.), ***Science in Athletics,*** Del Mar, Calif.: Academic Publishers, 1979.

Terauds, J. and Daniels, D.B. (Eds.),***Science in Gymnastics,*** Del Mar. Calif.: Academic Publishsers, 1979.

Terman, L.M. ***Genetic Studies of Genius.*** Stanford. Califo. Standford University Press, 1925.

Thomas, J.T. Keynote Address–The Army Looks at Biomechanics. In D. Bootzin and H.C. Muffey (Eds.), ***Biomechanics,*** New York: Plenum Press, 1969.

Thompson, C.W. ***Manual of Structural Kinesiology*** (7th ed.). St. Louis: C.V. Mosby Company, 1973.

Thurstone. L.I. ***Primary Mental Abilities,*** Chicago: University of Chicago Press, 1938.

Tichauer, E.R. Occupational Biomechanics and the Development of Work Tolerance. In P.V. Komi (Ed.), ***Biomechanics*** *V.A.* Baltimore: University Park Press, 1976.

Todon, J.I. Age Difference in the Generality and Speeviews of Motor Learning and Performance. Ph. D. Dissertation, University of California, Berkeley, California, 1972.

Tricker, R.A. R. and Tricker, B.S.K. ***The Science of Movement.*** New York: American Elsevier Publishing Company, Inc., 1967.

Van Gheluwe, B. A New Three–Dimensional Filming Technique Involving Simplified Alignment and Measurement Procedures. In R.C. Nelson and C.A. Morehouse (Eds.). ***Biomechanics IV.*** Baltimore: University Park Press, 1974.

Van Gheluwe, B. Computersied Three–Dimensional Cinematography for Any Arbitrary Camera Setup. In E. Asmussen and K. Jorgensen (Eds). ***Biomechanics VI.*** Baltimore: University Park Press, 1978.

Vincent, W.J. ***Statistics in Kinesiology*** Champaign, Human Kinetics, 1995.

Vorro, J.R. and Hobart, D. J. Multi–Image Stroboscopic Photographic Techniques for the Classroom and for Research. ***Journal of Health, Physical Education and Recreation,*** May 1973, pp. 63-64.

Vredenbregt, J. and Wartenweiler. J. (Eds). ***Biomechanics II.*** Baltimore: University Park Press, 1971.

Wainwright, S. Vogel, S. Denny, M.W., Eylers, J.P. Glasser, M.K. LaBarbera, M. Pearshall, G. W. and Clark, H.G. Letter: Concerning Words and People, ***Journal of Biomechanics,*** 1975, 8, 167.

Walters, C.E.Cooper, J.M. and Young, O. Kinesiology and Activity Analysis, In M.G. Scott (Ed.), ***Research Methods in Health, Physical Education and Recreation*** *(2nd ed.).* Washington D.C.: American Association for Health, Physical Education and Recreation, 1959.

Walters, C.E. and Partridge, M.J. An Elctromyographic Study of the Differential Action of the Abdominal Muscles. ***American Journal of Physical Medicine,*** 1957, 36, 259-268.

Walton, J.S. Close–Range Cine–Photogrammetry: Another Approach to Motion Analysis. In J. Terauds (Ed.), ***Science in Biomechanics Cinematography.*** Del Mar, Calif.: Academic Publishers, 1979.

Walton, J.S. and Kane, R. Interactive Computer Graphics: A New Coaching Aid. In E. Asmussen and K. Jorgenson (Eds.), ***Biomechanics VI-A,*** Baltimore: University Park Press, 1978.

Wartenweiler, J. Status Report on Biomechanics. In S. Cerquiglini, A. Venerando and J. Wartenweiler. J. Joki. E. and Hebbelinck, M (Eds.). ***Biomechanics III.*** Baltimore: University Park Press, 1973.

Wartenweiler, J. Joki, E. and Hebbelinck, M.(Eds.) ***Biomechanics.*** New York: S. Karger, 1968.

Washpurn, S.L. and Howells, F.C. ***Human Evolution and Culture.*** In the Evolution of Man, Vol.2 (S. Tax.Ed.). Chicago: University of Chicago Press,1960.

Wechsler, D. ***Intelligence Scale for Children.*** New York: Psychological Corporation, 1949.

Wells, K.F. ***Kinesiology.*** Philadelphia: W.B. Saunders Company, 1950.

Wells, K,F. William Skarstrom, M.D., Teacher in Kinesiology. ***Journal of Health, Physical Education and Recreation,*** Sept., 1968, 77-78.

Wells, K.F. and Luttengens, K. Kinesiology: ***Scientific Basis of Human Motion*** *(6th Ed.),* Philadelphia: W.B.

Wheeldon, D. Biomechanics: Space — Age Sports Medicines. ***The Physician and Sports Medicine,*** 1949.

Whittbe, M. and Harries, D. (Eds.) ***Biomechanical Measurement in Orthopaedic Practice.*** Oxford Clarendon Press, 1985.

Wickstrom, R. Developmental Kinesiology: Maturation of Basic Motor Patterns. In J. Wilmore and J. Keogh (Eds.), ***Exercise and Sports Sciences Reviews*** (Vol.3) Washington D.C.: American Asso. for Health, Physical Education and Recreation, 1971.

Widule, C.J. ***Analysis of Human Motion:*** Laboratory Experience, Experiments and Problems, Lajayette: Balt Publishers, 1974.

Williams, M. and Lissner, H.R. ***Biomechanics of Human Motion.*** Philadelphia: W.B. Saunders Co., 1962.

Winter, D.A. ***Biomechanics of Human Movement:*** New York, John Wiley and Sow, 1979.

Winter, D.A. Calculation and Interpretation of Mechanical Energy of Movement. In R.S. Hutton (Ed.) ***Exercise and Sport Sciences Reviews*** (Vol.6). Philadelphia: Frankin Institute Press, 1979.

Winter, D.A. Greenlaw, R.K. and Hobson, D.A. Television – Computer Analysis of Kinesiology of Human Gait. ***Computer and Biomedical Research,*** 1972, 5, 498-504.

Winter D.A., Sidwall, H.G. and Hobson, D.A. Measurement and Reduction of Noise in Kinesiology of Locomotion, ***Journal of Biomechanics,*** 1974, 7, 157-159.

Woltring, H.J. New Possibilities for Human Motion Studies by Real Time Light Spot Position Measurement. ***Biotelemetry,*** 1974 (3), 132-146.

Woltring, H.J.Calibration and Measurement in 3 – Dimensional Monitoring of Human Motion by Optoelectronic Means. ***Biotelemetry,*** 1975, 2 (3-4).

Wood, J.A. and Jennings, L.S. On the Use of Spline Functions for Date Smoothing. ***Journal of Biomechanics,*** 1979, 12, 477-479.

Wright, W.G. ***Muscle Function.*** New York: Paul B. Hocbir, Inc., 1928.

Zatziorsky, V.M. Studies of Motion and Motor Abilities of Sportsman, In R.C. Nelson and C.A. Morehouse (Eds.). ***Biomechanics IV,*** Baltimore: University Park Press, 1974.

Zebas, C.J. Kinesiology Reviews of Muscle Function For Physical Education Majors. ***Journal of Health, Physical Education and Recreation,*** May, 1977:60-61.

Zirnicke, R.F. Caldwell, G and Roberts, E.M Fitting Biomechanical Data With Cubic Spline Functions, ***Research Quarterly,*** 1976, 47, 9-19.

Zingg, W. The Role of Biomechanics in Sports Medicine. ***Athletic Training,*** 1975, 10 (2), 74-76.